Emperor of Enzymes

A Biography of Arthur Kornberg, Biochemist and Nobel Laureate

Emperor of Enzymes

A Biography of Arthur Kornberg, Biochemist and Nobel Laureate

Errol C. Friedberg
University of Texas Southwestern School of Medicine, USA

World Scientific

NEW JERSEY • LONDON • SINGAPORE • BEIJING • SHANGHAI • HONG KONG • TAIPEI • CHENNAI • TOKYO

Published by

World Scientific Publishing Co. Pte. Ltd.
5 Toh Tuck Link, Singapore 596224
USA office: 27 Warren Street, Suite 401-402, Hackensack, NJ 07601
UK office: 57 Shelton Street, Covent Garden, London WC2H 9HE

Library of Congress Cataloging-in-Publication Data
Names: Friedberg, Errol C., author.
Title: Emperor of enzymes : a biography of Arthur Kornberg, biochemist and
  Nobel laureate / Errol C. Friedberg.
Description: [Hackensack], New Jersey : World Scientific, 2016. |
  Includes bibliographical references and index.
Identifiers: LCCN 2015030393 | ISBN 9789814699808 (hardcover : alk. paper) |
  ISBN 9814699802 (hardcover : alk. paper) | ISBN 9789814699815 (pbk. : alk. paper) |
  ISBN 9814699810 (pbk. : alk. paper)
Subjects: LCSH: Kornberg, Arthur, 1918–2007--Health. | Biochemists--United States--Biography. |
  DNA polymerases. | Medicine--Research.
Classification: LCC QP606.D46 F75 2016 | DDC 572.8/6--dc23
LC record available at http://lccn.loc.gov/2015030393

British Library Cataloguing-in-Publication Data
A catalogue record for this book is available from the British Library.

Copyright © 2016 by World Scientific Publishing Co. Pte. Ltd.

All rights reserved. This book, or parts thereof, may not be reproduced in any form or by any means, electronic or mechanical, including photocopying, recording or any information storage and retrieval system now known or to be invented, without written permission from the publisher.

For photocopying of material in this volume, please pay a copying fee through the Copyright Clearance Center, Inc., 222 Rosewood Drive, Danvers, MA 01923, USA. In this case permission to photocopy is not required from the publisher.

This book is dedicated to Roger, Tom and Ken Kornberg, three sons of Arthur who both loved him deeply as a father and revered him greatly as a scientist

Foreword

In 1953, Watson and Crick proposed the double helical structure for the DNA molecule. Earlier work had shown that DNA is the carrier of hereditary information from one generation to the next. This opened a new field of study, the biochemistry of heredity. The master of that field was Arthur Kornberg, whose life is chronicled in this book.

Kornberg was a revolutionary. Before 1953, those interested in how heredity works had approached the problem as an abstract issue because the molecule that carried heredity was not known. They were geneticists, scientists who examined the behavior of mutant traits like eye color. These traits were the markers of heredity but told us nothing about the mechanisms of heredity. Kornberg recognized that heredity, like almost everything else to do with life, is at heart a chemical process and became committed to uncovering the nature of that chemistry. Errol Friedberg, a biochemist himself, tells us the story of how Kornberg achieved a deep understanding of the biochemistry of heredity. Because Kornberg was also a charismatic individual who not only did biochemistry but saw it as a lens though which all of biology could be known, the story of his life becomes a saga of the development of modern biology at large. Kornberg was a father figure who nourished the development of many others and together they played a central role in making molecular biology the science that it is today, one of the key sciences shaping the modern world.

Heredity is the transmission of characteristics from one generation to the next. It involves mainly duplication: A human makes another human, a fox makes another fox. If DNA carries the specificity that allows duplication of an organism, then understanding DNA duplication must be the key to heredity, and it was that process of duplication that Kornberg set out to understand.

As his career unfolded, Kornberg became a man of myth who Friedberg rightly calls the Emperor of Enzymes. Enzymes are the proteins that do the work of the body, and after Kornberg earned his medical degree, rather than treat patients, he became infatuated with the power of enzymes to create the chemicals of life. His own life's work became finding new enzymes and determining how they did their work. Life is mainly carbon, hydrogen and oxygen but Kornberg's focus became another, rarer, element of life, phosphorous. It is phosphorous, linked to oxygen in phosphate, that holds together the DNA molecule, and understanding the chemistry of phosphate was central to understanding how DNA is duplicated. When Kornberg found the enzyme that could make the linkage that holds together the DNA molecule, making the so-called phosphodiester bond, he opened a new chapter in biochemistry. Some years later, he was awarded his richly deserved Nobel Prize for this discovery. Kornberg memorably said: "You have to know the actors in order to understand the plot. And the actors are the enzymes. They are the mini-chemists, the devices by which a biological phenomenon takes place."

The great figures in scientific research are often not just insightful experimentalists, they are often also institution builders and trainers of the next generation. Kornberg took those roles to a special level. Starting in St. Louis, he aggregated around himself a group of the brightest young people and melded them into a cooperative juggernaut. Then he took most of them to Stanford where he set them up in a splendid new building and supported them as they developed individual careers of notable prominence. Virtually all of them stayed at Stanford for their careers, although any one could have had a position at almost any school he wished. Kornberg made mentorship a fine art.

Not only was Kornberg's mentorship of his faculty colleagues particularly effective, his mentorship of his own laboratory produced many notable investigators. I am particularly glad that Friedberg describes Kornberg's strict rules about how to organize a notebook. In that way, I am a student of Kornberg's because I learned from his colleague, Jerry Hurwitz, how to write a Kornbergian protocol, and to this day any graduate of Kornberg's laboratory could understand any of my experiments from my notebooks.

Duplicating DNA is a complicated process and Kornberg's discovery of a DNA polymerase was just the beginning of elucidating the process. He went on to find many more pieces of the puzzle and put each in its place. His iterative style of finding an assay, using it to discover an enzymatic activity, purifying the enzymes responsible and learning how they fit with other enzymes to make a biological process happen is largely a lost art. It was a very powerful style of research with many practitioners. But no one was better at it or more devoted to it than Arthur Kornberg. Friedberg has brought him to life for us, describing well the man I knew and deeply admired.

The saga of modern biology is a huge story that must be told in pieces. Errol Friedberg concentrates here on a key style of contribution as embodied in the work of its premier exponent. He has done history a great service.

David Baltimore

Nobel Laureate David Baltimore is President Emeritus of the
California Institute of Technology, and
Robert Andrews Millikan Professor of Biology

Preface

Arthur Kornberg is a household name to every biomedical scientist in the world. Born and raised in New York to a family that had to struggle to make financial ends meet, he had a prodigious intellect which was quickly recognized at school, where he was advanced by skipping several grades. Kornberg attended City College in New York before being admitted to the University of Rochester Medical School, from which he graduated in 1941. After a short stint in the US Navy he spent the years of World War II at the National Institutes of Health in Bethesda, Maryland, where he cultivated an abiding interest in biochemistry, especially the discipline of enzymology. While residing in Bethesda he married Sylvy Levy, a trained biochemist in her own right who also hailed from Rochester.

Postdoctoral stints in the laboratories of the Spanish biochemist Severo Ochoa at New York University and the Austrian-born Carl Cori at Washington University in St. Louis matured Kornberg's interest into sophisticated enzymological expertise that soon earned him an international reputation, so much so that in 1953, at the tender age of 35, he was offered and accepted the chairmanship of the Department of Microbiology at Washington University. While in St. Louis, Kornberg discovered an enzyme that faithfully copies a DNA template, an enzyme that he named DNA polymerase, which heralded a body of work that won him a Nobel Prize in 1959, an accolade memorably shared with his former mentor Severo Ochoa.

In that year Kornberg relocated to Palo Alto, California, to assume the chairmanship of a new Department of Biochemistry at Stanford University, taking with him the majority of the senior trainees in his

Washington University laboratory, all of whom went on to highly distinguished careers in biochemistry, molecular biology and genetics at Stanford, with a principal focus on deoxyribonucleic acid (DNA), a department widely hailed as the premier Department of Biochemistry in the USA — if not the world; an academic department that, in addition to Kornberg's own Nobel, spawned three more Nobel laureates from the ranks of its faculty and trainees*. The complex biochemistry of DNA replication endured as a primary focus of Kornberg's attention until the early 1990s, during which time he deciphered much of what we know today about genomic biochemistry. His contributions also paved the way to the generation of recombinant DNA technology by Paul Berg, one of the cadre of trainees who joined the Kornberg laboratory in St. Louis and one of the Nobel Laureates mentioned above.

While DNA replication in prokaryotic cells remained the primary focus for the majority of Kornberg's scientific career, in the early 1990s he shifted his attention to polyphosphate, a polymer found in plants, bacteria and animal cells and once a major topic of investigation by him and his wife Sylvy. Kornberg was convinced that polyphosphate was critical to the evolution of the early cellular organisms on Earth.

Kornberg and Sylvy, who tragically died of an untreatable neurological disorder in 1986, raised three sons, Roger, Tom and Ken. Following in the footsteps of his father, Roger Kornberg, a professor in the Department of Structural Biology at Stanford at the time of this writing collected a Nobel of his own in 2006. Tom Kornberg in turn, a highly accomplished cellist who would have made classical music a career choice if not for a debilitating physical handicap that abruptly terminated his career path as a musician, turned to molecular biology and made important contributions to our understanding of DNA replication through his discovery of additional DNA polymerases in the bacterium *E. coli*, ultimately acquiring a professorship at the San Francisco campus of the University of California, and Ken artfully married the passion for science that permeated the Kornberg household to his passion for architecture.

---

*Paul Berg, Nobel prize in Chemistry (1980); Randy Schekman and James (Jim) Rothman, Nobel prizes in Physiology or Medicine (2013).

A prolific and accomplished writer, Arthur Kornberg authored a number of leading texts on the topic of DNA replication. He also penned a veritable library of essays and commentaries, as well as a delightfully informative autobiography entitled *For the Love of Enzymes: The Odyssey of a Biochemist,* a literary contribution primarily focused on his career as a scientist, but which by design included scant contributions about his life outside the laboratory. This calculated oversight by Kornberg was a major stimulus to this author to more fully document his life.

Readers should be aware that this book makes no claim to trace the history of the long-standing and burgeoning field of DNA replication, an area of scientific endeavor to which many scientists contributed in addition to Arthur Kornberg. The book is intended to educate readers to the life and times of one of the great biochemists and enzymologists of the late 20th and early 21st centuries; one of the pioneers of the biology and chemistry of the genetic material nested in our chromosomes — with a particular focus on the multitude of enzymes that drive those feats of nature. This biography is also the story of a man who was forced to endure the torment of losing the love of his life and the mother of his three sons to the ravages of an untreatable neurological disorder — a loss amplified by the death of his second wife not too many years later.

Errol C. Friedberg
May 2015

Acknowledgments

First and foremost I wish to acknowledge Roger Kornberg for suggesting that I undertake the daunting task of writing a biography of his father, and for providing much helpful counsel during the course of my writing. His brothers Tom and Ken were also of invaluable assistance in many ways. I gratefully acknowledge the splendid Foreword to this book from David Baltimore, President Emeritus of the California Institute of Technology and Robert Andrews Millikan Professor of Biology. I thank the following individuals for interviews granted during my research: Robert (Buzz) Baldwin, Tania Baker, Paul Berg, Jack Griffith, Jerry Hurwitz, Dale Kaiser, David Korn, Carolyn Kornberg, Ken Kornberg, Tom Kornberg, Roger Kornberg, Bob Lehman, Randy Schekman, Eric Shooter, Jim Spudich, George Stark, Herb Tabor, Alan Waitz and Bill Wickner. Buzz Baldwin, Paul Berg, Dale Kaiser, Bob Lehman, Ken Kornberg, Roger Kornberg, Tom Kornberg, George Stark and Bill Wickner read drafts of the manuscript in various stages of completion. I am most grateful for their diligence and helpful advice. I also thank Lubert Stryer for information provided and Jamie Chermel for her incomparable logistical support. Last but no means least, thanks to Ms. Sook Cheng Lim and her staff at World Scientific Publishing for their dedication and patience in transforming hundreds of pages of typed text and multiple photographs to a handsomely illustrated and painstakingly edited book with a comprehensive index. Any remaining errors are mine.

In 1997 Arthur Kornberg gave a lengthy and informative interview to Sally Smith Hughes of the *Program in the History of the Biosciences and Biotechnology at the University of California*, which included an introduction by the late Joshua Lederberg. I borrowed heavily from this indispensable source. Kornberg's autobiography *For the Love of Enzymes: The Odyssey of a Biochemist* published in 1989 is another source from which I gathered much critical information. I recommend both of these sources to readers wanting to learn more about the life of Arthur Kornberg.

Contents

Foreword vii
Preface xi
Acknowledgments xv

Chapter 1	Growing Up in New York City, Medical School and the United States Navy	1
Chapter 2	Learning from the Masters — The NIH and the Ochoa Laboratory	17
Chapter 3	The Cori Laboratory — And a Return to the NIH	35
Chapter 4	Washington University	53
Chapter 5	The Lure of Nucleic Acid Enzymology	69
Chapter 6	The Discovery of DNA Polymerase	85
Chapter 7	Stanford University Medical School	97
Chapter 8	The Stanford Department of Biochemistry	131
Chapter 9	Life in the Test Tube?	159
Chapter 10	Like Father Like Sons	173
Chapter 11	Spores — A Brief Interlude	193
Chapter 12	DNA Replication — The Holy Grail	199
Chapter 13	Polyphosphate	225
Chapter 14	A Life of Writing	231

Chapter 15   In Support of Basic Science   255
Chapter 16   Commercial Ventures and the Founding of *DNAX*   267
Chapter 17   A Life Well Lived   285

Appendices   291

Index   297

CHAPTER ONE

Growing Up in New York City, Medical School and the United States Navy

The story of Arthur Kornberg's family is a saga that was repeated by a multitude of Jewish refugees around the turn of the 19th century. Kornberg's paternal grandfather David Lieb Kornberg was born and raised in Russia, where he died in 1918, the year that witnessed the end of World War I. Kornberg's paternal family name was originally Queller (sometimes spelled Kweller), a name that derives from Cuéllar, of Spanish Sephardic origin, but it was abandoned by his grandfather to avoid conscription to the Russian army. "The army draft was a fate no orthodox Jew could contemplate; to escape it he had taken the name of one Kornberg who had already done his military service," Arthur Kornberg explained in later years.[1] Arthur's paternal grandmother Bella, who bore the maiden name Krell, was born in Denmark. David Lieb and Bella were cousins. Bella's death preceded that of her husband, an event that prompted him to marry Bella's sister, ostensibly to provide for her — not an uncommon custom among Jewish families of that era. No children issued from this marriage. David Lieb lost his second wife to illness as well. His third wife bore him three children.

Arthur Kornberg's father Aaron Joseph (Joseph to those who knew him in the United States) was born in 1877. While in Europe he spent his working hours as a farm manager,[2] a skill of no obvious value in New York City, to which he moved in 1900 to escape the bleak

1

Arthur Kornberg's parents, Joseph and Lena.

prospects for Jews in the small towns that populated Eastern Europe. In New York Joseph met Lena Katz, who bore a not too dissimilar family history. The couple married in 1904. Joseph was 23 years old. Lena was a mere waif of 16.

A proficient tailor, a skill that he presumably acquired in Europe, Joseph resorted to walking the streets of New York with his sewing machine on his back in search of customers who could be persuaded to purchase hand-made items of clothing. Otherwise he worked in some poorly lit basement, laboring for 10 hours a day, 7 days a week — for a pittance. These grinding financial hardships notwithstanding Joseph managed to bring his parents and other members of his family to the United States from Europe — and later helped put Arthur through medical school. A deeply religious man, Joseph was fluent in eight languages, including Yiddish, Hebrew, English, Polish, Russian and German, and he taught himself to read and write English.

Arthur's parents lived in a Queens residence that boasted an outdoor toilet, a convenience then considered something of a luxury.[3] By the time that ailing health prompted him to seek a less strenuous occupation Joseph had accumulated sufficient financial resources to acquire a small hardware and house-furnishing store in the Bath Beach section of Brooklyn.[4,5] His handsomely printed stationary read:

Joseph Kornberg,
Homefurnishings, Paints, Electrical and Radio Supplies,
1727 Bath Avenue, Brooklyn.

The home kitchen and dining room were located behind the store; the bedrooms above it.[6] "We constantly teetered on the edge of poverty but somehow managed to stay out of debt," Kornberg wrote in his autobiography *For the Love of Enzymes: The Odyssey of a Biochemist.*[7] As a teenager young Arthur was expected to help relieve the family's precarious financial situation by waiting on customers in his father's store and maintaining the store inventory when not at school. "From age nine I helped out in my parents' house-furnishings and hardware store," he wrote, "a feeble and unprofitable business — and then clerked in other stores, particularly in men's furnishings (haberdashery) shops, during evenings and weekends throughout high school and college and for summers thereafter."[8]

Arthur Kornberg's eldest sibling Martin, a brother 13 years his senior, was a schoolteacher in New York who worked as a streetcar motorman during the summers to help meet financial ends. Arthur's other sibling Ella was nine years older than him. One of her roles was taking care of her younger brother, who apparently required more than a little watching over. "Both my parents and my sister Ella ----- recalled many incidents of my restlessness and abandon as a child, pulling away from them and getting lost," he wrote in later years.[9] Kornberg's wanderlust as a child and teenager was evident in his choice of grade and high schools well outside his Bath Beach neighborhood in Brooklyn.[10] He attended grade school in Brooklyn at a time when the most able students advanced by simply "skipping" grades. But as he had jumped two grades and was being considered for a third, his brother Martin intervened "out of concern for my social adjustment."[11]

As a child Kornberg never enjoyed experiences that might have aroused his interest in biology. "Why science?" he asked rhetorically. "I don't remember having any curiosity about my natural surroundings."[12] In later years, when he was working in the laboratory of the celebrated biochemistry team of Carl Cori and his wife Gerty at Washington University in St. Louis, MO (see Chapter 3), Gerty Cori once casually informed Arthur that her husband Carl had collected beetles and butterflies in his youth. When she asked Arthur what *he* had collected as a youngster his timid response was: matchbook covers! "They were the dominant flora in the Brooklyn streets where I played and in the subways where my father often risked being trampled as he stopped to add one more to my collection," Kornberg wrote.[13]

Like his boyhood Jewish friends Kornberg attended Hebrew school daily, as well as synagogue services on the high holy days. He also braved the exacting preparations required for a formal barmitzvah when he reached the age of 13. But he never cultivated a serious religious attachment and by the time he left home to attend medical school Kornberg could no longer indulge his parents by observing religious rituals. He found "the narrow views of my rabbinical teachers and the biblical accounts of miracles clearly at odds with the more interesting cosmology and biology I was learning at school. I found them meaningless and irrational," he wrote in an unpublished essay entitled "My Faith in Science."[14]

Kornberg spent his high school years at Abraham Lincoln High School in Brooklyn, an experience that he once described as being distinguished only by finishing three years ahead of schedule, but where he cultivated a strong interested in chemistry. Unlike Paul Berg, his future close friend and faculty colleague at Washington University, St. Louis and later at Stanford University, who attended Abraham Lincoln High School several years later, Kornberg did not document recollections of a teacher named Sophie Wolfe, who Berg described in the most inspirational terms in his own biography. "I recall nothing inspirational from teachers or courses except encouragement to get good marks," Kornberg wrote.[15] However, in

June 1968, when well entrenched as Chair of the Department of Biochemistry at Stanford University, he received a letter from Ms. Wolfe, stating: "I felt deeply flattered, gratified and honored when I received your reprints and very gracious, highly complimentary notes."[16]

Upon completing high school Kornberg attended City College in uptown Manhattan, despite the harsh reality that commuting from his home meant suffering three hours a day on the crowded subway. Originally founded as the Free Academy of the City of New York in 1847 by wealthy businessman and president of the Board of Education, Townsend Harris, the City College of New York was established to provide children of immigrants and the poor access to free higher education based strictly on academic merit. When it opened its doors on January 21, 1849, Horace Webster, the first president of the Free Academy stated: "The experiment is to be tried, whether the children of the people, the children of the whole people, can be educated; and whether an institution of the highest grade, can be successfully controlled by the popular will, not by the privileged few." City College thus became one of the nation's great democratic educational experiments.[17]

Even in its early years the Free Academy demonstrated remarkable tolerance for diversity, especially in comparison with the private universities in New York City.[18] But competition at City College among a large body of bright and highly motivated students was fierce in all subjects. "Only one or two A's were given in a class section of thirty," Kornberg wrote.[19] Kornberg, Paul Berg and Jerome Karle (another of Abraham Lincoln High School's esteemed products) all won Nobel Prizes, and each has a wing of the main Lincoln High School building graced with their name.

While at City College Kornberg maintained his high school interest in chemistry, but the prospects for employment in college teaching chemistry or in industry were bleak. In 1937, at age 19 and armed with a Bachelor of Science degree, he disconsolately found that there were no jobs to be had during the Great Depression.[20] So he set his sights on becoming a physician, one of 200 pre-medical students who applied to the University of Rochester medical school in 1937, only five of whom were accepted.[21] "I had chosen medicine because I was an avid student,"

Arthur with his parents outside their haberdashery store in New York.

he wrote, "and at age nineteen, with a Bachelor of Science degree from the City College of New York I welcomed the haven that medical school would provide for four more years. There were no better alternatives in 1937 in the depths of the depression."[22] With earnings of about $14 a week, supplemented by the combination of a New York State Regents Scholarship of $100 a year, some help from his parents — and frugal living, Kornberg saved enough money to see himself through medical school at the University of Rochester in upstate New York.

In medical school Kornberg was pleased to discover a relatively relaxed and non-competitive environment. He noted that "the curriculum was uncrowded, grades were not divulged and personal competition in a class of forty-four students was minimal."[23] "I enjoyed medical school, and I anticipated becoming an internist with academic connections," he wrote.[24] "The courses in anatomy and physiology presented fascinating images of the intricate organization and functions of the human body; in bacteriology and pathology we became acquainted with the morbid aberrations of horrid diseases. In classical medical-student fashion, I immediately succumbed to each of them — amyotropic lateral sclerosis, aortic aneurysm, lymphoma — only to be cured by preoccupations with still other diseases."[25] During his sojourn as a medical student in Rochester, Kornberg met his future wife Sylvy Levy, who gained a Master's degree in the Department of Biochemistry at the medical school, where she conducted research on body lipids. She was among the earliest investigators to use radioactive phosphate to trace the metabolism of phospholipids. Beginning in 1942 Sylvy worked at the National Cancer Institute in Bethesda, MD, where her research was focused on the synthesis of novel carcinogens and their effect on mice.[26]

When in Rochester, Kornberg maintained a brisk correspondence with his parents. His father composed occasional letters in Yiddish. But when retinal detachments largely destroyed his vision, his mother Lena took over the responsibility of written communication and attentively wrote to her son in English on a daily basis. To that end she heroically braved night school to improve her English diction. "At age 54, many years after she arrived in America, Lena, "with endearing enthusiasm and misspellings, first learned to read and write English," Kornberg wrote.[27] Lena also kept her son well supplied with cakes and strudels, all of which apparently survived the US mail to Rochester.

Following medical school Kornberg completed an internship in internal medicine at Strong Memorial Hospital, the university hospital at the University of Rochester Medical School, an experience that he found to be disappointingly bereft of any simulating intellectual content. "I was engrossed by the responsibility of coping with the complex problems of sick people," he related.

> I felt compassion for and a deep rapport with distressed patients, but despite an occasional diagnostic success, the lack of rigor and rational methods was discouraging to me. Signs and symptoms had to be promptly assigned a categorical niche, followed immediately by a pre-scribed treatment. There were few opportunities to think about the physiological basis of any disease.[28]

Those familiar with Arthur Kornberg's brilliant career as a biochemist might surmise that his introduction to the discipline of biochemistry was exhilarating. Not so. "My first encounter with biochemistry and the Bodansky text (referring to a biochemistry textbook written by Meyer Bodansky entitled *Introduction of Physiological Chemistry* and published in 1928) was in my course in medical school," he wrote. The focus of the text and the course was on the lysis of blood, urine and tissues and I found it utterly boring."[29]

> The dynamism of cellular energy exchanges and macromolecules was still unknown and the importance of enzymes had not penetrated my courses or textbooks. ----------. The idea of spending a significant fraction of my future days in the laboratory had no appeal.[30]

His early lack of interest in biomedical research notwithstanding, Kornberg was aware that the medical school at the University of Rochester granted a handful of student fellowships to take a year out of medical school in order to learn the fundamental elements of wet bench research. Kornberg hoped to obtain such an award from any of the numerous academic departments to which he applied. "I applied for and was eligible for several, but, even with my superior academic record, I failed to get any," he related.[31] "In those years, ethnic and religious barriers were formidable, even within the enlightened circle of academic science," he wrote.[32] Notwithstanding these disappointments, William S. McCann, Chairman of the Department of Medicine at the University of Rochester Medical School, persuaded a wealthy patient to endow a scholarship of which Kornberg was the first recipient. The scholarship yielded $100 for him to acquire reagents for his first research project, one described shortly.[33]

The religious barriers that Kornberg referred to surfaced by way of the subtle (sometimes not so subtle) anti-Semitic overtones then rampant in American medical schools — and presumably in other elements of higher education. At grade and high school in Brooklyn Kornberg was surrounded by a circle of Jewish friends and students and was blissfully unaware of prevailing anti-Semitic sentiments. The same was true at City College, the student body of which was then overwhelmingly Jewish. Then came the disappointment of being rejected by virtually all of the many medical schools to which he applied. Kornberg was one of just two of the many Jewish applicants admitted to the freshman class at the University of Rochester Medical School in 1937.

Kornberg felt strongly about the religious prejudice that he encountered as a Jew in an academic medical center in the early and mid-20th century. In the concluding chapter of his autobiography *For the Love of Enzymes: The Odyssey of a Biochemist*, a book about which more is revealed later (see Chapter 14) he devoted a substantial section to what he referred to as "the virus of anti-Semitism, a virus virulent around me early in my career."[34] "My first shock at Rochester was hearing anti-Semitic remarks and realizing how much more numerous they must be in my absence," Kornberg wrote.

> The worst was being denied academic awards and research opportunities because I was Jewish. I particularly coveted the pathology fellowship — a year's research and special training awarded to two students at the end of the course in pathology. The fellows were chosen by George H. Whipple, Nobel laureate, chairman of the Department of Pathology, dean of the school — and God of the Rochester medical universe. --------------
> I knew that my class performance was better than that of the students awarded the fellowships, and I later learned that I had ranked first in the class.[35]

In the course of a lengthy interview conducted in 1997 with Sally Smith Hughes, a historian of science from the Bancroft Library at the University of California, San Francisco, Kornberg revealed his views about anti-Semitism at the University of Rochester Medical School more explicitly. "My disappointment with the venerated don of medical education and science at the university, George Whipple, was well known to people at Rochester," he related.

> In June 1997 I was invited to give a lecture at the University of Rochester Medical School. Jay Stein, then the vice chancellor of the medical school informed me: "Your lecture will be attended by many people and the only auditorium we have that is large enough bears the name of someone you're not happy with. Is it all right with you?" I said, "Look, I know it's the Whipple Auditorium. Of course it's all right with me."[36]

"Wherever I go and recall the history of those times, anti-Semitism was rampant," Kornberg commented.

> It was true at UCLA; it was true at Stanford; it was true throughout the Midwest — Illinois, Wisconsin. I think Joshua Lederberg was the first Jewish assistant professor appointed at Wisconsin. But by that time he was a sensational star. A letter about Lederberg stated: "He is so brilliant, and he really doesn't have all the bad features of people of his race."[37]

In his 1997 interview with Sally Smith Hughes, Kornberg dwelt at considerable length and frankness on the issue of anti-Semitism. "Johns Hopkins was among the worst," he related.

> Barry Wood [a distinguished basic scientist and Harvard graduate who acquired his PhD at Stanford University in 1963] knew Johns Hopkins well. He had been a sensational All-American at Harvard. He became Professor of Medicine at Washington University [an institution that would prominently surface in Kornberg's own career; see Chapter 4] at a very young age. ---------- He went on to be chairman of Microbiology and vice chancellor of the medical school at Hopkins. At that time Hopkins had not offered an appointment to an outstanding young Jewish ophthalmologist, Bernie Becker, and Washington University did. Barry said to me, "You know, Hopkins has done it again. They've lost a most promising young star because of their anti-Semitic stance."

"When Barry appointed people to his department at Hopkins, he chose Dan Nathans, who was Jewish," Kornberg continued.

> On the other hand, Biochemistry, chaired by Al Lehninger, who wrote the most important textbook in biochemistry — and was a good friend of mine — didn't have a Jewish person in his department for twenty years, during a time when biochemistry was densely populated by Jews, either from Europe or home grown. Why not? He's dead now; I never approached him about it. I'm guessing that he aspired to be part of a social set and a country club that didn't admit Jews and extended this discrimination in his department.[38]

"I think it is important to be reminded that, like some viruses, anti-Semitism is endemic," Kornberg related.

> For that reason I want my students and my children, to whom anti-Semitism is completely foreign, to know that as recently as in my lifetime, it could be severe and ugly. I could flesh out more experiences like mine. Jewish colleagues who have written autobiographies and have had the opportunity to speak out have not done so.[39]

In later years Kornberg once reminded a former colleague at the University of Rochester Medical School of Whipple's anti-Semitic sentiments. "Let's not beat around the bush," he declared. "Whipple was anti-Semitic and it figured in his judgments." To which his friend offered the rejoinder: "You know, you're right. But he didn't like Italians either!"

As mentioned earlier, though conscious of his religious heritage, Kornberg was never a dedicated practicing Jew. But he maintained a sentimental tie to the state of Israel for much of his adult life. He visited the country as a guest of one or another research establishment on multiple occasions, and in 1965 was elected an Honorary Fellow of the Weizmann Institute. In later years he was a member of a group from Stanford University and the University of California at Berkeley who met with US Senator Alan Cranston to help develop a Marshall Plan for the Middle East. In a letter to Cranston in February 1974 he informed the Senator of a plan

> that is, in my estimation enormously attractive. It promises imaginative solutions to these problems: productive employment for refugees in Jordan and Lebanon, ----- and an impact on neighboring Arab countries; generalization of cheap energy for the long term in oil-poor regions, utilization of precious minerals in the desert, and peaceful use of nuclear energy for excavation of an Eilat-Dead Sea canal.[40]

In the same year Kornberg accepted a plea from the President of Israel Ephraim Katzir to help in the development of a new Center for the Advanced Study of Science and Technology.

****

Biomedical research ultimately found its way into Kornberg's heart and mind when as a medical student he noticed that he was mildly jaundiced and his blood bilirubin levels were modestly elevated. After tracking down a cohort of seven other medical students with the same symptoms and signs, he examined the medical records of 17 former patients who had recovered from infectious hepatitis, as well as those of healthy subjects and patients with unrelated disease entities. More detailed examination of the group of seven jaundiced students revealed that they were unable to eliminate bilirubin normally following an intravenous load of the compound. These observations formed the basis of Kornberg's first scientific paper, entitled "Latent Liver Disease in Persons Recovering From Catarrhal Jaundice and in Otherwise Normal Medical Students, as Revealed by the Bilirubin Excretion Test," published in the *Journal of Clinical Investigation* in May 1942. Kornberg's careful scrutiny of the literature revealed that Augustin Nicolas Gilbert (a French physician) had first described this syndrome, which now bears his name.[41]

Plagued by the large number of cases of jaundice among recruits inoculated with yellow fever vaccine, the Medical Corps of the US Army and Navy took notice of Kornberg's publication and dispatched a team of physicians to the University of Rochester Medical School for advice. The team first sought out George Whipple of course. Imagine Whipple's astonishment (and presumed chagrin) when the group asked to see Arthur Kornberg, the other "authority" on jaundice at Rochester Medical School![42]

"Looking back, I realized that I enjoyed collecting data," Kornberg wrote.[43] He continued collecting bilirubin measurements during his internship year at Strong Memorial Hospital, and when he enlisted in the US Navy during World War II he began setting up to do further analyses in the small sickbay of a Navy ship that he joined in August 1942. Being in the Navy was the first time in his adult life that Kornberg found free time on his hands. Liberated from the need to work for money to support himself while simultaneously studying, he viewed the experience as the vacation he'd never had! A man of definite opinions and unabashed confidence, Kornberg not infrequently bumped heads with the captain of the naval vessel to which he was assigned.

Arthur Kornberg in the 1930s.

Kornberg was once afforded the opportunity of performing a hemitonsillectomy on one of the ship's crew under the supervision of a trained ear nose and throat surgeon while his ship was in port in St. Petersburg, Florida. The surgeon competently removed one tonsil while explaining the surgical procedure to Kornberg, who was then assigned the removal of the remaining organ. Kornberg failed to appropriately anesthetize the tonsil, "delivering most of the drug to the patient's throat," causing the unfortunate sailor to lunge "straight up out of the chair and let out an ear-splitting shriek!" The next day the mortified Kornberg tried to avoid meeting the sailor on deck, "but he followed me to say: 'Doc, the side you did feels great but the one that butcher did is killing me.'"[44]

By extraordinary coincidence, Bernard Davis, a colleague at the NIH (where, as presently related, Kornberg was reassigned from his maritime stint) was once stopped on a New York street by a sailor, who, upon recognizing the Army Medical Corps insignia of a caduceus surrounded by an anchor that adorned Davis's uniform, asked him whether he by any chance knew a navy doctor named Kornberg. When informed by Davis that he indeed did, the sailor admiringly stated: "Great surgeon!"[45]

Rolla Dyer, Director of the NIH from 1942–1948 took particular note of Kornberg's publication about cases of Gilbert's Syndrome and engineered Kornberg's transfer from sea duty to perform research at the National Institutes of Health (NIH) in Bethesda, MD — a rare and coveted war time assignment, especially for someone with no significant research experience. Kornberg's transfer did not at all displease the captain of his ship, who had been repeatedly exasperated by the young doctor's ignorance of military discipline — not to mention Kornberg's indifference to his supreme authority! A heated discussion about some medical administrative matter usually ended with his superior stating: "I'm the captain!" To which Kornberg apparently often replied: "But I'm the doctor!"[46]

References

1. Kornberg A. (1991) *For the Love of Enzymes: The Odyssey of a Biochemist.* Harvard University Press, p. 3.
2. Ibid.
3. ECF interview with Ken Kornberg, Jan. 2014.
4. Arthur Kornberg, http://en.wikipedia.org/wiki/Arthur_Kornberg
5. Kornberg A. (1991) *For the Love of Enzymes: The Odyssey of a Biochemist.* Harvard University Press, p. 3.
6. Ibid.
7. Ibid.
8. Kornberg A. (2002) *The Golden Helix: Inside Biotech Ventures.* University Science Books, p. 21.
9. Ibid.
10. Ibid.
11. Ibid., p. 19.
12. Ibid., p. 21.
13. Kornberg A. (1991) *For the Love of Enzymes: The Odyssey of a Biochemist.* Harvard University Press, p. 3.
14. Kornberg A. "My Faith in Science," unpublished essay, Arthur Kornberg Archives, Special Collections, Stanford University Libraries.
15. Kornberg A. (1991) *For the Love of Enzymes: The Odyssey of a Biochemist.* Harvard University Press, p. 2.

16. Letter to Arthur Kornberg from Sophie Wolfe, June 24, 1968. Arthur Kornberg Archives, reproduced with permission from the Department of Special Collections, Stanford University Libraries.
17. The City College of New York, About: Our History, www.ccny.cuny.edu/about/history.cfm
18. Ibid.
19. Kornberg A. (1991) *For the Love of Enzymes: The Odyssey of a Biochemist.* Harvard University Press, p. 4.
20. Ibid., p. 2.
21. Ibid., p. 4.
22. Ibid., p. 3.
23. Ibid., p. 4.
24. Kornberg A. (2002) *The Golden Helix: Inside Biotech Ventures.* University Science Books, p. 21.
25. Ibid.
26. Kornberg A. (1991) *For the Love of Enzymes: The Odyssey of a Biochemist.* Harvard University Press, p. 173.
27. Ibid., p. 4.
28. Kornberg A. (2002) *The Golden Helix: Inside Biotech Ventures*, University Science Books, pp. 21–22.
29. Kornberg A. (2002) How I Became a Biochemist. *Life* **53**: 185–186.
30. Kornberg A. (1991) *For the Love of Enzymes: The Odyssey of a Biochemist.* Harvard University Press, p. 6.
31. Kornberg A. (2002) *The Golden Helix: Inside Biotech Ventures.* University Science Books, p. 22.
32. Kornberg A. (1991) *For the Love of Enzymes: The Odyssey of a Biochemist*, Harvard University Press, p. 6.
33. Ibid., p. 311.
34. Ibid., p. 310.
35. Ibid., pp. 310, 311.
36. AK Interview with Sally Smith Hughes. Program in the History of the Biosciences and Biotechnology, Biochemistry at Stanford, Biotechnology at DNAX, 1997, p. 146.
37. Ibid.
38. Ibid.
39. Hargittai I. (2002) Arthur Kornberg. In: *Candid Science, II. Conversations with Famous Biomedical Scientists.* Imperial College Press, p. 53.

40. Letter from Arthur Kornberg to Alan Cranston, Feb. 11, 1974. Reproduced with permission from the Department of Special Collections, Stanford University Libraries.
41. Augustin Nicolas Gilbert, http:/en.wikipedia.org/wiki/Augustin_Nicolas_Gilbert
42. Kornberg A. (1991) *For the Love of Enzymes: The Odyssey of a Biochemist.* Harvard University Press, p. 6.
43. Kornberg A. (1989) Never a Dull Enzyme. *Ann Rev Biochem* **58**: 1–31.
44. Kornberg A. (1991) *For the Love of Enzymes: The Odyssey of a Biochemist,* Harvard University Press. pp. 6, 7.
45. Ibid., p. 8.
46. Ibid., p. 7.

CHAPTER TWO

Learning from the Masters — The NIH and the Ochoa Laboratory

The NIH traces its roots to 1887, when a one-room laboratory was created within the Marine Hospital Service (MHS), predecessor agency to the US Public Health Service (PHS). The MHS had been established in 1798 to provide for the medical care of merchant seamen. A clerk in the Treasury Department collected 20 cents per month from the wages of each seaman to cover costs at a series of contract hospitals. In the 1880s the MHS had been charged by Congress with examining passengers on arriving ships for clinical signs of infectious diseases in order to prevent epidemics, especially of cholera and yellow fever. Some years later the MHS authorized Joseph J. Kinyoun, a young physician trained in modern bacteriological methods, to set up a one-room laboratory in the Marine Hospital at Stapleton, Staten Island, New York. Kinyoun referred to this facility as a "laboratory of hygiene."[1] In due course the Hygienic Laboratory, as it came to be called, was moved to Washington, D.C., and was eventually recognized in law when Congress authorized funds for construction of a new building in which the laboratory could investigate infectious and contagious diseases as well as matters pertaining to the public health. In 1930 the name of the Hygienic Laboratory was changed to National Institute (singular) of Health (NIH).[2] Over the years the term "institute" became pluralized as the NIH grew to become a physically massive and leading bastion of biomedical research.

Arthur in his Navy uniform, with his father and an unidentified child.

The Nutrition Laboratory at the NIH to which Kornberg was assigned in the fall of 1942 as a commissioned officer in the U.S. Public Health Service was established by Joseph Goldberger, a physician and epidemiologist in the U.S. Public Health Service. A strong advocate of scientific and social recognition concerning the links between poverty and disease, Goldberger was nominated five times for the Nobel Prize for his work on the etiology of pellagra.[3] He was among the first to recognize that vitamin deficiency can cause disease, and in tracking the missing vitamin in the diets of pellagra patients he emerged as one of the greatest of the so-called "vitamin hunters" (a phrase borrowed from Paul de Kruif's *Microbe Hunters*). When discovering the cause of pellagra Goldberger stepped on a number of medical toes since his research demonstrated that diet and not germs (the prevailing medical theory then) caused the disease.[4] He in turn trained William Henry Sebrell, a physician, researcher in nutrition, administrator and educator, who became recognized for his work on riboflavin (vitamin B2), pantothenic acid (vitamin B5), folic acid (vitamin B9) and thiamin (vitamin B1). As chief of the Nutrition Laboratory, Sebrell was Kornberg's nominal boss.

Kornberg's first assigned research project as a nutritionist was to determine why rats fed a synthetic diet containing a sulfa drug developed a severe blood disorder in a few weeks and died, and why a stock animal ration or inclusion of a yeast or liver supplement to the purified diet was effective in preventing and curing the disease. "It seemed that we were dealing more with an induced dietary deficiency than a drug toxicity," he related.[5] Following in the footsteps of other "vitamin hunters" who succeeded in isolating folic acid and making it available to the public, Kornberg and his colleagues demonstrated that an induced deficiency of this vitamin was indeed responsible for the sulfa drug effect.

Kornberg was elated to be working with rats rather than caring for patients. With hundreds of inbred rats he could devise month-long experiments and get definitive results. "The serious economic disadvantages of choosing a career in research over the lucrative practice of medicine never entered my mind — a strange indifference, considering the poverty of my childhood and the many years of working, skimping, and borrowing to get through school," he wrote years later. "Nor were the consequences of this choice considered when housing and maintaining a family made increasing demands on a modest, fixed salary."[6] Over a period of little more than two years Kornberg amassed an impressive list of publications, most co-authored with Sebrell. "But," he wrote, "I always felt that I had come to the field of nutrition research in its twilight, decades too long to share the excitement and adventures of the early vitamin hunters who had solved riddles of diseases that had plagued the world for centuries."[7]

In the early 1940s Kornberg, then not even a fledgling biochemist, was singularly impressed by a seminar at the NIH presented by the geneticist Edward Tatum, in which Tatum described seminal work that he and George Beadle had carried out at Stanford University with mutants of the bread mold *Neurospora crassa*, work that demonstrated that each enzyme is encoded by a single gene. "I was greatly inspired by that," Kornberg related. "It didn't affect my [own] work, but much more than most people in nutrition [at that time] I was absorbing [new] things."[8]

Tatum was then a renowned biochemist and geneticist who had moved to Stanford University from the University of Wisconsin, where he received his PhD in 1934. In 1937 he began his celebrated collaboration with Beadle. The experiments that so enthralled young Kornberg, published in 1941, led to the famous "one gene, one enzyme" hypothesis that correctly postulated a direct link between genes and enzymes, a fact now taken for granted by students at high school, if not earlier. In 1958 Tatum and Beadle shared half the Nobel Prize for this contribution to science. The other half of the prize was won by Joshua Lederberg for his discoveries concerning genetic recombination and the organization of the genetic material of bacteria.[9] Lederberg was soon to become a faculty colleague of Kornberg's at Stanford University.

After listening to Tatum's lecture Kornberg disconsolately conceded that he knew even less about genetics than about biochemistry! But he became increasingly more intrigued by "a new breed of hunters tracking down the metabolic enzymes, the intricate machines that used vitamins to catalyze the combustion of sugar in yeast and muscle to generate the energy for their growth and work."[10] "I vividly recall my excitement in reading for the first time about these enzymes and about the miracle molecule called ATP," he wrote.[11] As he widened his sphere of scientific interests Kornberg was stimulated by reading about enzymes, coenzymes and the high energy compound adenosine triphosphate (ATP), the major source of energy during all cellular metabolism, and he came to the conclusion that "in trying to encompass medicine, agriculture, economics, psychology and anthropology, nutrition had become more a social and political activity and less a science."[12] He abandoned the animal nutrition laboratory when he realized that "enzymes are the vital force in biology, the sites of vitamin actions, and the means for a better understanding of life as chemistry."[13]

During the war years the NIH housed two young and accomplished biochemists, Bernard (Bernie) Horecker and Leon Heppel. Horecker began his training in enzymology in 1936 as a graduate student in the laboratory of Thorfin Rusten Hogness, a physical chemist in the Department of Chemistry at the University of Chicago. Horecker's initial project involved studying succinic dehydrogenase from beef heart. He subsequently collaborated with Erwin Hass from Otto Warburg's laboratory in the search for an enzyme that could catalyze the reduction of cytochrome c by reduced nicotinamide adenine dinucleotide phosphate

(NADP). This marked the beginning of Horecker's lifelong involvement with the pentose phosphate pathway.[14] When assigned to the NIH, Horecker was given the task of developing a method to determine the carbon monoxide content of the blood of Navy pilots returning from combat missions.[15] He was also required to investigate how the insecticide DDT killed cockroaches![16]

Leon Heppel was born to a poor Mormon family in Granger, Utah, received his doctorate in biochemistry from the University of California, Berkeley in 1937, and like Kornberg he received his medical degree from the University of Rochester in 1941 and also completed his medical internship at Strong Memorial Hospital. His research efforts during this period revealed that Na^+ and K^+ ions were capable of crossing animal membranes, contrary to the entrenched belief that the lipid cell membrane prevented the passage of hydrophilic metals.[17] Heppel and Kornberg were classmates and friends, and Kornberg's assignment to the NIH from ship duty was likely considerably facilitated by Heppel's strong recommendation to Rolla Dyer. While at the NIH during the war years Heppel was required by the Navy to carry out toxicology research. After the war, in collaboration with his longtime colleague Russell Hilmoe, Heppel focused on enzymes that hydrolyzed RNA, particularly spleen phosphodiesterase. The nature of the products formed and the phosphodiester bond hydrolyzed by this enzyme were elucidated by Heppel during a sabbatical period at the Molteno Institute in Cambridge, England (1953) in collaboration with Roy Markham and John D. Smith.[18]

Kornberg teamed up with Horecker to begin learning enzymology. The pair studied the effect of cyanide on the succinic dehydrogenase system. Cyanide had previously been reported to inhibit enzymes containing a heme group, with the exception of cytochrome c. However, Kornberg and Horecker established that cyanide did in fact react with cytochrome c and concluded that previous investigators had failed to perceive this interaction because the shift in the absorption maximum was too small to be detected by visual examination, but could be revealed by spectrophotometry.[19]

"Bernie introduced me to succinoxidase, cytochrome c, and the Model DU spectrophotometer, a new and powerful analytical biochemical

Bernard (Bernie) Horecker, Kornberg's mentor at the NIH.

Leon Heppel. Another early mentor.

tool," Kornberg related.[20] "Our probings with cytochrome c and its oxidation did not bring us to our goal, which was to discover the major source of ATP generation," he wrote.[21] Still, the collaboration yielded a paper published in the *Journal of Biological Chemistry* in 1946 entitled "The Cytochrome c-Cyanide Complex," Kornberg's first publication in contemporary biochemistry.

Kornberg's early interest in vitamins and the realization that they in fact served as cocatalysts (or coenzymes) — detachable working parts of an enzyme — to perform detailed chemical operations, led him to an increasing focus on enzymes in general. "I vividly recall my excitement in reading for the first time about enzymes, coenzymes and the high energy compound adenosine triphosphate (ATP) in papers by Otto Warburg, Otto Meyerhof, Carl Cori, Herman Kalckar and Fritz Lipmann," (all leading biochemists of the day), he wrote. "From my medical school training I had learned nothing about these people or the things they were investigating."[22]

While at the NIH Kornberg renewed his acquaintance with Sylvy Ruth Levy, who Kornberg first met when the two resided in Rochester, NY. Sylvy had acquired a Master's degree in biochemistry from the University of Rochester, where her main research focus was on body lipids under the tutelage of Walter R. Bloor, once described as "the leader of a small group of biochemists, largely his own students, who were working in the field of lipids."[23] In 1941 Sylvy published a paper in the *Journal of Biological Chemistry* on the occurrence and rate of turnover of sphingomyelin in tumors. While in Bethesda she occupied a staff position at the National Cancer Institute, where she investigated the synthesis of novel carcinogens and their effects on mice. Arthur and Sylvy married in 1943. In subsequent years they enlarged their family with the births of three sons, Roger, Tom and Ken, born in 1947, 1948 and 1950, respectively. As her family grew in size Sylvy ceased her laboratory work, being mainly occupied with raising her three sons. Still, she found time to edit books at home for Interscience Publishers.[24]

****

At war's end Kornberg was determined to acquire more advanced training as a biochemist, and he petitioned Sebrell to allow him to immerse himself in research environments where he could learn more about enzymes. He set his sights on learning in depth about ATP and the enzymes and coenzymes associated with intermediary metabolism, a topic that embraces the collective intracellular processes by which nutritive material is metabolized in living cells, with the concomitant generation of energy. The study of intermediary metabolism defined cutting edge biochemical research for much of the 19th and 20th centuries, yielding a formidable cadre of outstanding biochemists in the United States and Europe before it was subsumed by the discipline of molecular biology, with the biochemical focus shifting to understanding genes and how they work.

When he ultimately convinced the NIH to sponsor his work in laboratories outside the NIH, Kornberg first considered joining David Keilin's laboratory in Cambridge, England. "But I was told that his lab wasn't functioning well as an aftermath of the war," he related. "One of the American refugees from Cambridge, David Green, invited me to his lab, but I had friends who said David was a difficult character." Meanwhile Kornberg had become aware of a young Spaniard, Severo Ochoa, who was working at the New York University Medical School. "Bernie Horecker and I [had] read his papers and found that he was doing the kind of enzymology we'd love to do," he related. "So I applied to work with him. He had nobody else so he took me! I was still in uniform."[25]

> Perhaps the most important feature that characterizes the few who emerge from the pack of trained, intelligent, motivated scientists is the capacity to withstand distractions and disappointments in life at home and at large, in institutional duties and politics, and in lack of resources and recognition, and also to resist the temptations of fame and fortune. Severo Ochoa was one of these few

Kornberg wrote in 2001.[26] Born and raised in Spain, Ochoa entered medical school at the University of Madrid in 1922,[27] where he hoped to work under the tutelage of the renowned Spanish neuroscientist and Nobel Laureate Ramon Cajal. Disappointed to discover that Cajal had already retired, Ochoa acquired exposure to biochemical research when in the summer of 1927 he was appointed assistant to one of Spain's most distinguished physiologists, Juan Negrin, at the University of Madrid. Ochoa also enjoyed a sojourn in Scotland working with the physiologist/biochemist Noel Paton at the University of Glasgow. The year 1929 found him in Germany in the laboratory of the famous German biochemist and Nobel Laureate Otto Meyerhof, who together with Gustav Embden and Jakob Karol Parnas discovered the metabolic reaction sequence during glycolysis by which glucose is converted to pyruvic acid with the attendant generation of ATP and NADH,[28] a reaction popularized as the Embden–Meyerhof–Parnas pathway.

Under Meyerhof's direction Ochoa settled into the biochemistry and physiology of muscle, all the while acquiring a scientific outlook strongly influenced by his mentor.[29] In 1938 he joined R. A. Peters at Oxford, where he developed a keen interest in oxidative metabolism, prompting another geographical move, this time to the United States to work with Carl and Gerty Cori at Washington University in St. Louis, Missouri. In the early 1940s Ochoa acquired a faculty position at the New York University Medical School, rising to the position of Professor and Chairman of the Department of Biochemistry. During this period he also acquired American citizenship.

Over the years Ochoa's research mainly dealt with enzymatic processes in biological oxidation and synthesis, and in energy transfer, work that among his other studies contributed much to understanding basic steps in the metabolism of carbohydrates and fatty acids and the utilization of carbon dioxide in metabolism. A polished biochemist, Ochoa's eclectic research interests included the biological functions of vitamin B, oxidative phosphorylation, the reductive carboxylation of ketoglutaric and pyruvic acids, the photochemical reduction of pyridine nucleotides in photosynthesis, and the function of citrate synthase (so-called condensing enzyme), an enzyme in the Krebs citric acid cycle that catalyzes the condensation of oxaloacetate, water and acetyl-CoA, generating citrate and coenzyme A.

"When I went to work with Severo Ochoa in 1946 enzymology was utterly new to me," Kornberg related.[30] "That was my introduction to genuine biochemistry and enzymology, and from then on that was all I was interested in and wanted to do." "I was very fortunate that Ochoa was willing to take me, a complete novice in all aspects of biochemistry [under his wing]," he wrote. "I hardly knew the difference between ADP and ATP. I knew even less about enzymes."[31]

Regardless of the fact that he had grown up and been educated in New York, and Sylvy had often visited the city from Rochester, the couple found themselves to be relative strangers to the city. Adding to their discomfort was the grim reality that they were initially essentially homeless, since vacant apartments in New York immediately after the war were practically non-existent. Providentially, they secured an apartment occupied by the Danish biochemist Herman Kalckar and his wife, who were about to return to their native Denmark — but not until mid-February of 1946. For a period of about six weeks the Kornbergs moved from one hotel to another, "including the best and the sleaziest."[32] While Arthur spent his days (and not infrequently nights) in the Ochoa laboratory, it was not uncommon for Sylvy to check out of a hotel in the morning and spend much if not an entire day hopefully searching for a new one at the Officers Service Club, an organization that assisted ex-servicemen. When Kalckar and his wife finally left New York, Arthur and Sylvy moved into their premises on West 123 Street.

Kornberg joined Ochoa's laboratory a day after Christmas of 1945 totally committed to learning more biochemistry — and finding the mother lode of ATP. Ochoa occupied borrowed space in an old laboratory in the Department of Biochemistry at New York University. When he entered the laboratory Kornberg found six fresh pig hearts waiting for him on his assigned bench. His mission was to purify heart muscle aconitase — his first solo stab at enzyme purification. The enzyme was thought to consist of two components since *cis*-aconitic acid was an intermediate. But aconitase turned out to be a single enzyme. Purifying the enzyme did not come easy to the young biochemical novice. "'This was my first stab at enzyme purification and after a few months' work I failed to make any significant progress," Kornberg wrote in later years.[33] Regardless of these initial setbacks "this introduction to enzymology was exhilarating," Kornberg stated.[34] "Aside from the fascination of seeing an enzyme in action, the pace of the work was breathtaking."[35] By coupling aconitase action to isocitrate dehydrogenase, spectrophotometric assays could be performed in a few minutes and multiple ideas could be tested and discarded in the course of a day. Late evenings were occupied preparing protocols for the following day. "What a contrast with the tedious pace of nutritional experiments on rats,"

Kornberg related.[36] "In my work on aconitase I learned the philosophy and practice of enzyme purification."[37]

From the very first assay Kornberg initiated a lifelong love affair with enzymes. "I have always been in awe of enzymes, even the so-called dull ones," he wrote. "With such affection, purifying an enzyme is always rewarding. I doubt that any effort in purifying an enzyme was ever wasted."[38] Such was the intensity of Kornberg's passion that it could move him to lyrical literary efforts. To him the process of enzyme purification

> often seemed like the ascent of an unchartered mountain: The logistics resembled supplying successively higher base camps; protein fatalities and confusing contaminants resembled the adventure of unexpected storms and hardships. Gratifying views along the way fed the anticipation of what would be seen from the top. The ultimate reward of a pure enzyme was tantamount to the unobstructed and commanding view from the summit. Beyond the grand vista and thrill of being there first, there was no need for descent, but rather the prospect of even more inviting mountains, each with the promise of even grander views.[39]

In subsequent work Kornberg joined Ochoa and his graduate student Alan Mahler in studies on a liver enzyme called malate dehydrogenase, which reversibly catalyzes the oxidation of malate to oxaloacetate, a reaction intrinsic to multiple metabolic pathways, including the citric acid (Krebs) cycle. Graduate student Mahler became Kornberg's "indefatigable and devoted tutor"[40] Having always been the youngest in his class, Kornberg admitted frank surprise to discover that he was so far behind someone four years his junior!

Working in a biochemistry laboratory is in many ways akin to cooking, in terms of the care and skills required. But as those who have immersed themselves in home kitchens well know, accidents sometimes happen! Shortly before completing his stint in Ochoa's laboratory, Kornberg, with the help of others, undertook a large-scale purification of malate dehydrogenase, starting with several hundred pigeons that had "donated" their livers! After several weeks of toil, often into the wee hours, and having reached the final step of the enzyme purification protocol, one evening Kornberg had the chilling experience of watching the entire enzyme preparation accidentally spill on the floor. He was both shocked and mortified. Ochoa was in the laboratory and "by the time I got home about an hour later he had called Sylvy," Kornberg recounted. "I had been so terribly upset that Ochoa was concerned about my safety while returning home."[41]

Upon returning to the laboratory the following morning (presumably having lost much if not an entire night's sleep) Kornberg dejectedly glanced at the last saved supernatant fraction from the previous hellish night's work, which he had fortuitously saved and stored in the refrigerator, with no particular plan for its use. In so doing he noticed that the supernatant had taken on a slightly turbid appearance, suggesting that some of its contents had formed insoluble precipitates. On a whim he was prompted to harvest the precipitated material by centrifugation, redissolve it and test it for malic acid enzyme activity. "Holy Toledo!" he exclaimed. "This fraction had the bulk of the enzyme activity and was several-fold purer than the best of our previous preparations."[42] This purification step (sans the cylinder breakage) became part of the published purification procedure.

During his six months in Ochoa's laboratory Kornberg keenly felt his inadequacy in organic and physical chemistry, an educational deficiency suffered by most if not all biochemists and molecular biologists formally trained as physicians rather than as scientists. Accordingly, he was prompted to enroll in summer courses at nearby Columbia University, an experience that among other things resurfaced old fears of tests and examinations. "I found it strange in taking these optional courses that the dread of exams returned, along with the familiar nightmares of entering the examination room totally unprepared," he wrote in later years in his autobiography, *For the Love of Enzymes: The Odyssey of a Biochemist.*

> And the grades, which I would share only with my wife, provoked the same chills and apprehensions I knew in my intensely competitive school and college years. Despite taking the courses I never gained the fluent and unaccented command of these basic disciplines that I might have enjoyed had they been part of an early curriculum,[43]

a limitation shared by many who rose from the ranks of physician-turned-biochemist.

Kornberg's sojourn in Ochoa's laboratory was intended to be no longer than six months. However, the NIH acceded to his plea to extend this to a full year. He described this experience as "one of the happiest and most exhilarating in my life. Never had my learning curve been so sharply exponential and sustained."[44]

> I learned something new every day. I wasn't doing anything important experimentally, but I was acquiring scientific language and attitudes. Ochoa was a very impressive person, with his enthusiasm, optimism, and his breadth of vision. ---------- That year with Ochoa was absolutely great; every day mattered. ------------------------ I had confidence that my association with Ochoa gave me access to and familiarity with those who were identified as being at the forefront of biochemistry. I felt that I knew what they were doing and that I was capable of doing something along those lines. I was confident that if I worked hard and did things carefully I would find something worthwhile.[45]

Kornberg never forgot the debt he believed he owed to Ochoa. "Ochoa gave me the opportunity to learn about enzymes," Kornberg told interviewer Sally Smith Hughes in 1997.

> "There was a sublime air about him ---- an intense interest in science, not seeming to worry about what would happen the next day, an enthusiasm for the results, even if they weren't that interesting. Did I learn from him? I don't know. Could I have learned something like that? I don't think so. [Perhaps] you're born with it ------- and you are reinforced by seeing it practiced by someone who you admire. ----- The other thing I learned was that I didn't think that Ochoa was that much smarter than I. I thought I could achieve what he had."[46]

Kornberg's reverence for the esteemed Spanish-born biochemist emerges unmistakably between the lines of his writing, leaving an unassailable impression of the special place that Ochoa found in his heart and mind. In ending personal correspondence with Ochoa he was not immune to replacing the well-worn term "Sincerely" with "Love."

Living in New York City was another experience not to be missed. Arthur and Sylvy reveled in discovering the theater, music, and museums with which New York is so famously populated. Kornberg's work with Ochoa yielded four publications in the *Journal of Biological Chemistry*, three of which were published in 1948. That year marked the publication of a further five contributions to the journal, two of which took the form of letters to the editor authored by Kornberg alone based on work carried out on his return to the NIH: one on the participation of inorganic pyrophosphate in the enzymatic synthesis of diphosphopyridine nucleotide and another on nucleotide pyrophosphatase and triphosphopyridine nucleotide structure.

In the early 1970s Kornberg celebrated Ochoa's 70th birthday by helping orchestrate an International Symposium organized by the Autonomous Universities of Barcelona and Madrid and the University of Houston, with the collaboration of Stanford University and the Roche Institute of Molecular Biology. The symposium convened in both Barcelona and Madrid in September 1975. In his recent book *Crucible of Science: The Story of the Cori Laboratory*, author John Exton quotes Kornberg as describing the event "as a party, the likes of which has not been seen in scientific circles before or since." The celebrations even

Severo Ochoa, a prominent figure in Kornberg's scientific development.

included a visit to the illustrious Spanish artist Salvador Dali in his museum in the town of Figueres, in Catalonia, Spain, Dali's birthplace.[47]

The symposium embraced six colloquia in fields in which Ochoa had made special contributions: energy metabolism, lipids and saccharides, regulation, nucleic acids and the genetic code, protein biosynthesis and cell biology. A year later this memorable gesture was immortalized in a substantial volume entitled *Reflections on Biochemistry — In Honour of Severo Ochoa*, edited by Kornberg, Bernie Horecker and two Spanish scientists, L. Cornudella and J. Oro.[48] In the preface the editors commented:

> It would have been conventional and relatively simple to prepare a *Festschrift* of these papers. Instead, the participants attempted something more difficult. They wrote essays reflecting on the development of a subject, a concept, or an approach to biochemistry.

Kornberg's own contribution to the commemorative volume began with a sentence often repeated in his writing: "I have never known a dull enzyme." Making special mention of enzymes that he worked on with Ochoa, he continued:

> succinic dehydrogenase, aconitase, isocitric dehydrogenase and malic enzyme; these were brief encounters during the first year I met enzymes, but my recollections of them are still vivid and tinged with pleasure. There is said to be a monogamous relationship with enzymes: "one man, one enzyme." It is even more true that some enzymes support the careers of many men. Despite a happy marriage for 20 years to DNA polymerase [an enzyme discovered by Kornberg in later years that propelled his career to the front ranks of biochemistry; see later chapters], I feel impelled in this age of frank revelation to expose flirtations, affairs and fantasies with other enzymes.[49]

The commemorative volume notably included a homage to Ochoa by Salvador Dali, who poetically wrote:

> Even though I am not a scientist, I must confess that the scientific events are the only ones that guide constantly my imagination, at the same time that they illustrate the poetic intuitions of traditional philosophers, to the point of the blinding beauty of certain mathematical structures,

specially those of the polytopes, and above all those sublime moments of abstraction, which 'seen' through the electron microscope, appear as viruses or regular polyhedral form, confirming what Plato said: 'God always makes Geometry'.[50]

As related in a later chapter, Kornberg shared the 1959 Nobel Prize in physiology or medicine with his former mentor, an occasion that one presumes had special significance for him. When Ochoa died in 1993 at the age of 88, Kornberg wrote a moving eulogy. He described the man as a courtly, charming El Greco-like figure who, with his wife Carmen, was a most gracious host, and he commented on the couple's appreciation of music, fine food, travel, and good company.[51]

Ochoa played host to a legion of postdoctoral fellows, students and sabbatical guests who came from every corner of the world and left to become leaders in science. His many medals, diplomas, citations and honorary degrees are displayed in a handsome museum founded by a distinguished disciple, Santiago Grisiola, and located in Valencia near a street that bears his name, as do most cities in Spain. Don Severo Ochoa was a name and a face nearly as well known to the Spanish cab driver and housewife as that of King Juan Carlos.[52]

References

1. A Short History of the National Institutes of Health, history.nih.gov/exhibits/history/index.html
2. Ibid.
3. Joseph Goldberger, en.wikipedia.org/wiki/Joseph_Goldberger
4. Kornberg A. (1991) *For the Love of Enzymes: The Odyssey of a Biochemist*. Harvard University Press, p. 8.
5. Ibid., p. 17.
6. Kornberg A. (2002) *The Golden Helix: Inside Biotech Ventures*. University Science Books, p. 22.
7. Kornberg A. (1991) *For the Love of Enzymes: The Odyssey of a Biochemist*. Harvard University Press, p. 8.
8. AK Interview with Sally Smith Hughes, Program in the History of the Biosciences and Biotechnology, Biochemistry at Stanford, Biotechnology at DNAX, Arthur Kornberg, 1997, p. 20.

9. The Nobel Prize in Physiology or Medicine 1958, www.nobelprize.org/nobel_prizes/medicine/laureates/1958/
10. Kornberg A. (1991) *For the Love of Enzymes: The Odyssey of a Biochemist.* Harvard University Press, p. 29.
11. Ibid., p. 30.
12. Ibid., p. 29.
13. Ibid., p. 30.
14. Kresge N, Simoni RD, Hill RL. (2005) Contributions to Elucidating the Pentose Phosphate Pathway. *J Biol Chem* **280**: e26.
15. Ibid.
16. Kornberg A. (1991) *For the Love of Enzymes: The Odyssey of a Biochemist.* Harvard University Press, p. 43.
17. Hurwitz J. (2010) Leon A. Heppel (1912–2010), *ASBMB Today.* http://www.asbmb.org/asbmbtoday/201007/Retrospective/Heppel
18. Ibid.
19. Kresge N, Simoni RD, Hill RL. (2005) Bernard L. Horecker's Contributions to Elucidating the Pentose Phosphate Pathway. *J Biol Chem* **280**: e26.
20. Kornberg A. (1989) Never a Dull Enzyme. *Ann Rev Biochem* **58**: 1–31.
21. Kornberg A. (1991) *For the Love of Enzymes: The Odyssey of a Biochemist.* Harvard University Press, p. 47.
22. Ibid., p. 30.
23. Sperry WM. (1966) Walter R. Bloor. *The Clinical Chemist* **12**: 897–899.
24. Kornberg A. (1991) *For the Love of Enzymes: The Odyssey of a Biochemist.* Harvard University Press, p. 173.
25. AK Interview with Sally Smith Hughes, Program in the History of the Biosciences and Biotechnology, Biochemistry at Stanford, Biotechnology at DNAX, Arthur Kornberg, 1997, p. 10.
26. Kornberg A. (2001) Remembering Our Teachers. *J Biol Chem* **276**: 3–11.
27. Severo Ochoa — Biographical, http://www.nobelprize.org/nobel_prizes/medicine/laureates/1959/ochoa-bio.html
28. *Glycolysis*, http://en.wikipedia.org/wiki/Glycolysis
29. Severo Ochoa — Biographical, http://www.nobelprize.org/nobel_prizes/medicine/laureates/1959/ochoa-bio.html
30. AK Interview with Sally Smith Hughes, Program in the History of the Biosciences and Biotechnology, Biochemistry at Stanford, Biotechnology at DNAX, Arthur Kornberg, 1997, p. 10.
31. Kornberg A. (2001) Remembering Our Teachers. *J Biol Chem* **276**: 3–11.

32. Kornberg A. (1991) *For the Love of Enzymes: The Odyssey of a Biochemist.* Harvard University Press, p. 50.
33. Kornberg A, Horecker BL, Cornudella L, Oro J. eds. (1976) For the Love of Enzymes. In: *Reflections on Biochemistry in Honour of Severo Ochoa.* Pergamon Press, p. 244.
34. Ibid.
35. Kornberg A. (1989) Never a Dull Enzyme. *Ann Rev Biochem* **58**: 1–31.
36. Ibid.
37. Kornberg A. (1991) *For the Love of Enzymes: The Odyssey of a Biochemist.* Harvard University Press, p. 51.
38. Kornberg A. (1989) Never a Dull Enzyme. *Ann Rev Biochem* **58**: 1–31.
39. Kornberg A. (1991) *For the Love of Enzymes: The Odyssey of a Biochemist.* Harvard University Press, p. 52.
40. Ibid., p. 57.
41. Ibid., p. 58.
42. Ibid., p. 57.
43. Ibid.
44. Ibid., p. 50.
45. AK Interview with Sally Smith Hughes, Program in the History of the Biosciences and Biotechnology, Biochemistry at Stanford, Biotechnology at DNAX, Arthur Kornberg, 1997, p. 11.
46. Ibid., pp. 11, 12.
47. Exton JH. (2013) *Crucible of Science: The Story of the Cori Laboratory.* Oxford University Press, p. 38.
48. Kornberg A, Horecker BL, Cornudella L, Oro J. eds. (1976) *Reflections on Biochemistry In Honor of Severo Ochoa.* Pergamon Press.
49. Ibid., p. 243.
50. Ibid., p. 445.
51. Exton JH. (2013) *Crucible of Science: The Story of the Cori Laboratory.* Oxford University Press, p. 37.
52. Kornberg A. (1993) Obituary, Severo Ochoa (1905–1993). *Nature* **366**: 408.

CHAPTER THREE

The Cori Laboratory — And a Return to the NIH

In 1947 Kornberg joined another prestigious laboratory for six months of further postdoctoral training, one operated by the husband and wife team of Carl and Gerty Cori (nee Radnitz) in the Department of Pharmacology at Washington University in St. Louis, an institution to which he was destined to return in a considerably more senior capacity not too many years hence.

Among the most celebrated biochemists of their time, Carl Cori and Gerty Radnitz hailed from Austrian families.[1] Carl grew up in Trieste, where his father was director of the Marine Biological Station.[2] As a youth, he often accompanied his father on field trips to collect marine samples. "Trieste was a fascinating city in which to grow up," Carl wrote in later years. "The early contact in school with different language groups developed in me an immunity against racial propaganda."[3,4]

Prior to his involvement with the Marine Biological Station Cori's father had studied medicine — and later earned a PhD in zoology. "It would have been unusual for me to go in a different direction," Cori wrote in an autobiographical article entitled "The Call of Science." "Rejection of the values of one's parents was not as prevalent then and family tradition was still a strong influence."[5]

With the outbreak of World War I Cori's family moved to Prague, where Carl entered medical school and where he met Gerty Radnitz. During the war he was drafted into the Austro-Hungarian Army and served in the ski corps. Carl and Gerty graduated from medical school in 1920 and married that year.⁶

Gerty Radnitz, a Jewess, was the daughter of a chemist who devised a method for refining sugar and became a successful manager of sugar refineries. Her mother, a cultured woman, was a friend of the influential novelist Franz Kafka. The oldest of three sisters, Gerty was educated at home until the age of 10, when she attended a private school.⁷

The many difficulties of life in post-World War II Europe, complicated by Gerty's Jewish heritage (not to mention the fact that she was a woman aspiring to become a scientist in an almost exclusively male-dominated profession) prompted the couple to think seriously about emigrating to the United States. No offers followed an initial search for research appointments, prompting Carl to accept a position with Otto Loewi in the Department of Pharmacology at the University of Graz in Vienna. Notwithstanding the undeniable satisfaction of working with Loewi (a celebrated biochemist who discovered acetylcholine and earned a Nobel Prize in Physiology or Medicine in 1936), Carl found the atmosphere at Graz increasingly uninviting. Living conditions were generally poor, and he was irked by the requirement to prove his Aryan descent in order to obtain employment at the university. Once again the couple turned their attention to America, this time with a promising outcome. In 1922 job offers for both Coris came from the State Institute for the Study of Malignant Disease (later called the Roswell Park Cancer Institute) in Buffalo, New York.⁸

In 1931 Carl Cori was offered and accepted the chairmanship of the Department of Pharmacology at Washington University School of Medicine in St. Louis. His only previous academic experience was as an adjunct assistant professor at the University of Buffalo for one year. Gerty in turn was offered a research post in the same department — at a salary a fraction that of her husband's. In 1942 Cori received a joint appointment as professor of biochemistry at Washington University. Four years later he became head of that department.⁹

The Coris' most notable contributions to biochemistry embraced discoveries that elucidated the pathway of glycogen breakdown in animal cells and the enzymatic basis of its regulation. Their groundbreaking research into the enzyme-catalyzed chemical reactions of carbohydrate metabolism elevated the medical school at Washington University to the front ranks and won the Coris Nobel Prizes in Physiology or Medicine in 1947. Sadly, the excitement of receiving this coveted award was tarnished by the discovery that Gerty had developed myelosclerosis — an incurable and fatal illness. She died in 1957.[10] "She was a very volatile and inspiring person, a fascinating person, truly a great scientist," Kornberg said of Gerty Cori.[11] Carl shares a star with Gerty on the St. Louis Walk of Fame.[12,]* In 1960 Carl Cori married Anne Fitz-Gerald Jones, a woman with whom he shared a number of interests, including archeology, art and literature. Cori retired from Washington University in 1966 and he and his wife moved to Boston, where Cori was appointed visiting professor of Biological Chemistry at Harvard Medical School and maintained a laboratory at the Massachusetts General Hospital.

Carl and Gerty Cori in their Washington University laboratory.

---

*The St. Louis Walk of Fame honors notable people from St. Louis, Missouri, who made contributions to the culture of the United States. As of April 2014, the walk consisted of 137 brass stars and bronze plaques, set into the sidewalks of Delmar Boulevard, each containing an inductee's name and a summary of his or her accomplishments.

The Coris left a rich legacy to the world of biochemistry that notably included training successive generations of outstanding biochemists. No fewer than six future Nobel Laureates (Christian de Duve, Luis F. Leloir, Edwin G. Krebs, Arthur Kornberg, Severo Ochoa and Earl W. Sutherland) passed through their laboratory at Washington University at one time or another.[13] In September 2004 the American Chemical Society dedicated the research of Carl and Gerty Cori on the metabolism of carbohydrates at Washington University School of Medicine a National Historic Chemical Landmark. A plaque commemorating the event reads:

> Beginning in the 1920s, Carl and Gerty Cori conducted a series of pioneering studies that led to our current understanding of the metabolism of sugars. They elucidated the "Cori cycle," the process by which the body reversibly converts glucose and glycogen, the polymeric storage form of this sugar. They isolated and purified many of the enzymes involved in glucose metabolism. The work of the Coris advanced understanding of glycogen breakdown in cells and of metabolic regulation. Building on their work, others developed improved techniques to control diabetes. The Coris were awarded Nobel Prizes in 1947.[14]

Kornberg said of his mentor: "He had an extraordinary intellect and breadth of knowledge and was awesome that way. And yet he was so supportive of what I was doing that I was encouraged by his respect and confidence in me."[15]

In the 19th century it had been shown that respiratory processes consume oxygen and generate carbon dioxide: so-called *aerobic respiration*. However, reactions such as alcoholic fermentation in yeast cells can also sustain cellular functions via oxygen-independent (*anaerobic*) processes, but far less efficiently. A yeast cell in air (i.e., in the presence of oxygen) consumes only a twentieth as much glucose for its growth as it does during anaerobic fermentation. When a molecule of glucose is rearranged to yield two molecules of pyruvic acid the latter compound undergoes a number of consecutive enzyme-catalyzed reactions that convert it to carbon dioxide and water, yielding adenosine triphosphate (ATP) in the process. During aerobic respiration cells are able to generate and store about 20 times as much energy in the form of ATP for each glucose molecule consumed compared to that generated during anaerobic respiration.

ATP was discovered by the German chemist Karl Lohmann, and its structure established some years later. Achieving the biosynthesis of ATP in the test tube requires a complex enzymatic system resident in respiring tissues rich in mitochondria; tissues such as muscle, liver, kidney and brain. An enormous amount of ATP is required to convert the energy in food to the forms we need to live and work on a daily basis. Kornberg calculated that the average daily intake of 2,500 calories translates to a turnover of about 180 kilograms (~400 pounds) of ATP. Since the cellular content of ATP in humans is only 50 grams (about one-tenth of a pound), the cycling of ATP — its synthesis from ADP (adenosine diphosphate) plus inorganic phosphate, and its subsequent breakdown — must occur approximately 4,000 times a day. In 1945, the mechanisms responsible for the great majority of this ATP synthesis were unknown.[16]

Having acquired the basic skills essential to addressing biochemical problems in intermediary metabolism during his sojourn with Severo Ochoa, in the Cori laboratory Kornberg turned his attention to the burning question: exactly how is the bonanza of ATP generated [and stored] during aerobic respiration? "How does ATP store energy?" he wondered. "How is the energy used in the manufacture of even more complex entities, such as protein and DNA — and how is chemical energy converted to mechanical energy for movement?"[17]

In the Cori laboratory Kornberg was initially assigned to work with John Taylor, a junior faculty member in the Department of Pharmacology, with whom he was given the task of crystallizing lactic dehydrogenase, an enzyme that interconverts pyruvic and lactic acids, the production of the latter being a crucial step during the anaerobic metabolism of glucose. But Kornberg had little interest in this project, which clearly had nothing to do with the synthesis of ATP. In fact, he openly rebelled at continuing — an early, though not necessarily the earliest harbinger of the strong personality underlying his placid charm and patience. He had come to the Cori laboratory to solve the major problem in biochemistry, the mechanism of aerobic respiration and the generation of ATP — not to crystallize proteins! Nor was he in the least daunted by the challenge he had set himself. "It did not discourage me that Cori, Ochoa, Kalckar, and Lipmann, each of whom had contributed so much to the recognition of aerobic phosphorylation, had given up working on this problem," he wrote.[18]

With remarkable tolerance, perhaps born of his early recognition of his young postdoctoral fellow's obvious scientific talent, Cori suggested that Kornberg instead join a young Swedish member of the laboratory, Olov Lindberg, who was pursuing an observation reported by Severo Ochoa when he had worked with the Coris about six years earlier. Ochoa had discovered that liver mitochondria generated inorganic pyrophosphate (PPi), a previously unknown cellular constituent. The chemical energy trapped in pyrophosphate is in fact comparable to that trapped in ATP generated by the fusion of ADP with phosphate (Pi).[19] The mechanism of the origin of inorganic pyrophosphate was also unknown. But Kornberg's foray into this challenging biochemical arena did not bear the fruit he had hoped for. "We learned little about aerobic phosphorylation and nothing about the source of inorganic pyrophosphate," he wrote disconsolately.[20] "We had only a few ideas and even fewer rabbits on which to test them."

His gloom was in no way lightened by what Kornberg described as "the misery of a St. Louis summer, with no air conditioning" that ground experiments to a halt for weeks on end because of the unmanageable room temperatures. "This marked the end of my search for the source of ATP," he wrote. "It had been doomed from the start because I was committed to finding discrete soluble enzymes that linked the synthesis of ATP to aerobic metabolism. They don't exist that way. Instead these enzymes are firmly embedded in the walls of tiny intracellular organs called mitochondria."[21] "Those interested in pursuing these enzymes would be compelled to explore the intricacies of the organized mitochondrial enzyme system. I was not among them," he emphatically stated.[22]

But in the midst of this gloom a ray of sunshine that was in due course to progress to more brilliant illumination emerged. While attempting to enhance the levels of respiration and coupled aerobic phosphorylation by kidney mitochondria Kornberg and Lindberg observed a powerful stimulation by the coenzyme nicotinamide adenine dinucleotide (NAD) and discovered that the effect could be traced to the mononucleotide adenosine monophosphate (AMP) generated by NAD hydrolysis. The pair also succeeded in isolating a novel enzyme that they dubbed diphosphopyridine nucleotide pyrophosphatase that cleaved NAD yielding AMP and nicotinamide-ribose phosphate (NRP).

$$\text{Nicotinamide-ribose-P-P-ribose-adenine} + H_2O$$
$$\rightarrow \text{Nicotinamide-ribose-P} + AMP$$

Tom Kornberg points out the naming of this enzyme would seem illogical and confusing unless it is understood that NAD was initially called diphosphopyridinenucleotide and that it was renamed subsequent to its initial discovery and characterization. "Arthur objected to and was annoyed by this decision made by a nomenclature committee, and his annoyance never abated," Tom related.[23]

****

Laboratories in premier research institutions are never left empty for long. Predictably, when Kornberg returned to the NIH in the fall of 1947 he found his former laboratory space fully occupied. But his space needs were reasonably satisfied when Bernie Horecker provided him a bench in his own laboratory. Intently grooming independent research careers, around this time Horecker and another young biochemist, Leon Heppel, were flirting with the notion of leaving the NIH for greener pastures. In an effort to forestall the potential loss of two fine scientists Kornberg petitioned Sebrell about initiating a new research group at the NIH to be called the *Enzyme Section*, which would include Heppel, Horecker and himself. Sebrell agreed and Kornberg, now chief of a section (albeit a small one) had a laboratory to call his very own.

Much of the planned research of the group in the NIH *Enzyme Section* depended heavily on assays that utilized NAD and NADP. There were then no vendors of such biochemical reagents, so Kornberg and his colleagues set about isolating substantial quantities of these compounds from sheep liver using a method devised by Otto Warburg. This resulted in them acquiring the world's largest supply of NADP. When Warburg once visited their laboratory they were able to gift him with 25 milligrams of that coenzyme, a huge amount by any standard![24]

When in 1947 he returned to the NIH (which then consisted of just six small buildings with a primary focus on infectious diseases),[25] Kornberg continued to investigate the properties of diphosphopyridine nucleotide pyrophosphatase, the rabbit kidney enzyme that he and Lindberg had discovered in the Cori laboratory. Disconcertingly, these studies were plagued by an unwelcome affinity of the enzyme for aggregates of nondescript cellular material. Extracts of lamb, pig, or beef kidneys produced no better results. Once again good fortune intervened (as it not infrequently does for those with prepared minds) when Kornberg learned from two veteran NIH biochemists, Oliver Lowry and Sidney Colowick, about potatoes as a source of an enzyme with properties similar to those of diphosphopyridine nucleotide pyrophosphatase.

Oliver Lowry was another member of the distinguished cadre of American biochemists in the early 20th century. In the 1950s Lowry developed methods for isolating, preparing, weighing and chemically studying single nerve cells and subcellular particles. He also pioneered freeze-drying methods to preserve cells and invented a microbalance that could register less than a millionth of a gram. Using such microtechniques, Lowry performed biochemical studies on minute regions of the brain. Among his many contributions to biochemistry, Lowry is best known for the development of an assay used in the determination of protein concentrations — which typically bears his name. As of 2014, his 1951 paper in the *Journal of Biological Chemistry* describing the protein assay was the most-highly cited paper of all time, with more than 305,000 citations![26]

In 1947, Washington University invited Lowry to head its Department of Pharmacology, this despite the fact that he had never taken a course in pharmacology and his research was only tangentially related to that field. Nevertheless, Lowry served as the department head for the next 29 years. From 1955 to 1958 he also served as dean of the Washington University School of Medicine, an institution that was soon to play a major role in Kornberg's career.

Sidney Colowick in turn was Carl Cori's first graduate student and earned his PhD in 1942. Along with the Danish biochemist Herman Kalckar, he discovered myokinase, now known as adenyl kinase. This discovery proved to be important in understanding transphosphorylation reactions in yeast and animal cells. Colowick's interest then turned to the conversion of glucose to polysaccharides, and he did important work on the formation of glycogen from glucose using purified enzymes. Colowick also carried out studies on hexokinase, which led to his eventual crystallization of the enzyme in 1961.[27]

Concerned about variations in enzyme levels from different potato strains, Kornberg systematically examined extracts of multiple strains. He eventually grew weary of this rigorous approach and simply visited a large food market in Bethesda where he acquired samples "from each of a dozen nameless 100 pound sacks"[28] and returned the next day to buy the sack of potatoes that tested (as opposed to tasted) the best! After extensive purification of potato nucleotide pyrophosphatase Kornberg discovered that the enzyme also supported cleavage of another coenzyme, nicotinamide adenine dinucleotide phosphate (NADP), a compound chemically related to NAD, but possessing an extra phosphate group. The precise location of this phosphate in the compound had previously defied identification by others because of the harsh chemical treatments used. Kornberg demonstrated that the extra phosphate was attached to the second carbon of the ribose sugar in the ARP portion of NADP.[29]

Though not especially interesting to Kornberg at the time, diphosphopyridine nucleotide pyrophosphatase was to have momentous consequences for his future as a biochemist. In his autobiography he wrote that one of his greatest thrills transpired on a day in July 1948 when he and his wife Sylvy returned from a vacation to the Gaspe Peninsula in Canada. For several weeks before embarking on this vacation he had been working on experiments that he required for inclusion in a manuscript describing the properties of potato nucleotide pyrophosphatase — with little special interest. "These preparative and descriptive experiments seemed dull and my enthusiasm for them had diminished," he wrote.[30] He began to consider how he might gainfully utilize the nicotinamide-ribose phosphate (NRP) generated as a product of the potato enzyme's cleavage of NAD. "Might having this novel compound give me the opportunity to discover the mechanism of *synthesis* of the NAD coenzyme?" he wondered. Kornberg set about exploring this notion and resourcefully uncovered evidence for the reverse biochemical reaction. He established an absolute requirement for both NRP and ATP and showed that in addition to NAD, inorganic pyrophosphate (PPi) was generated during the reaction, another important biochemical observation, since it provided the first clue of the natural origin of PPi in living cells.

In 1948 Kornberg shared this exciting news in a letter to Cori:

> "During the past 2 months I have been working with a system which mediates the reaction: nicotinamide mononucleotide + ATP <-> DPN + inorganic pyrophosphate," he wrote. "It would seem from these results that inorganic pyro is a more 'physiologic' substance than we thought and I am anxious to have your thoughts."[31]

In due course biochemists came to the crucial realization that the basic mechanism of the biosynthesis of the coenzyme that Kornberg elucidated is recapitulated in the biosynthesis of proteins, carbohydrates and nucleic acids. The building blocks of a protein are amino acids. Accordingly, each amino acid reacts with ATP, releasing PPi, following which the activated amino acid is incorporated into a particular protein sequence. To generate fats and steroids, acetic acid units are activated by ATP in a similar way, also generating pyrophosphate. During the synthesis of carbohydrates and the phospholipids of membranes, building blocks react with analogues of ATP in which A is replaced by U or C, but the reaction mechanism follows the same pattern, with the attendant release of PPi. "In the ensuing years, the mechanism of nucleotidyl transfer from a nucleoside triphosphate for the biosynthesis of coenzymes was discovered again and again in the biosynthesis of proteins, lipids, carbohydrates, and nucleic acids," Kornberg noted.[32] As presently recounted, the observation that this biosynthetic pattern is reiterated during the synthesis of the nucleic acids DNA and RNA from nucleotide building blocks particularly piqued Kornberg's interest and ultimately encouraged him to forsake the world of intermediary metabolism and to enter that of nucleic acid enzymology, with a specific focus on the biochemistry of deoxyribonucleic acid (DNA), the mother lode of genetics.[33]

****

Having now acquired a robust mass of revealing biochemical data, Kornberg decided that the time was apposite to announce his findings in the biochemical literature. When, during the spring of 1949, he entered the United States Marine Hospital for the long delayed repair of bilateral inguinal hernias, he brought with him a huge briefcase crammed with data for the manuscripts that would describe the work of the previous year; to wit (i) the isolation of nucleotide pyrophosphatase (ii) the reversible enzymatic synthesis of NAD and inorganic pyrophosphate, (iii) the reversible synthesis of FAD and (iv) the enzymatic synthesis of NADP.

When shown to his room in the hospital Kornberg noted the absence of a desk and snapped irately at the nurse! "How can I spend the rest of today and the two or three days after surgery without a place to spread my notebooks and write the papers?" he grumbled. At this point the nurse (well experienced in dealing with difficult young officers) facetiously asked Kornberg whether he'd noticed a military uniform on a cart in the hall. "It belonged to a young medical officer who didn't survive his 'minor surgery,'" she announced. "I didn't die on the operating table," Kornberg joked. "But I also didn't miss having a desk during the painful postoperative days."[34]

Four back-to-back papers completed the following summer were published in the February 1, 1950 issue of the revered *Journal of Biological Chemistry*. This body of work earned Kornberg the prestigious Paul Lewis (now called the Pfizer) Award in Enzymology at the tender age of 33, for which he was nominated by none less than Otto Myerhof, who won the Nobel Prize in Medicine and Physiology in 1922 for his discovery of the relationship between the consumption of oxygen and the metabolism of lactic acid in muscle. As mentioned in the previous chapter, together with Gustav Embden and Jakob Parnas, Myerhof discovered glycolysis, the metabolic pathway that converts glucose to pyruvate, with the concomitant generation of ATP, commonly referred to as the Embden–Meyerhof–Parnas pathway.[35]

Kornberg's receipt of the Paul Lewis Award was no trivial token of recognition. Previous awardees included Van Potter, Albert Lehninger, Henry Lardy and Britton Chance, all distinguished American biochemists.[36] The award gained Kornberg widespread visibility and prominence. Among the many congratulatory letters he received was one from the renowned Danish biochemist Herman Kalckar, who wrote "your brilliant work on pyrophosphatase is the most original and important one in enzyme chemistry during the past 10 years."[37]

Arthur Kornberg's career as a biochemist was now clearly in the ascendancy. Besides the accolade of being chief of the *Enzyme Section* funded directly by the NIH, he was gaining the admiring attention of the international biochemistry community. Then too, and crucial to his future success, his laboratory had begun to attract bright young postdoctoral fellows. "Among them in 1952 were four future stars: Bruce Ames, Paul Berg, Edward Korn and Gordon Tomkins," Kornberg wrote. "Had Korn eventually joined me, as Berg did, we might have produced a paper by Korn, Berg and Kornberg," he playfully noted! Those who knew him well have commented on Kornberg's highly developed sense of humor — and his penchant for off-color jokes![38]

Enzymology now defined Arthur Kornberg's scientific career, and he waxed lyrically about these crucial cellular constituents at every opportunity. As late as 1997 he told interviewer Sally Smith Hughes:

> you have to know the actors in order to understand the plot. And the actors are the enzymes. They are the mini-chemists, the devices by which a biological phenomenon takes place, whether it is the legendary question of alcohol fermentation — how the juice of a grape generates a fine wine or champagne — that bedeviled people for over a century, or how a firefly comes to luminescence. Applying that same reductive approach to other phenomena, I believe we can get to the core of biologic questions — by finding the actors, and the actors are the enzymes.[39]

In his autobiography *For the Love of Enzymes: The Odyssey of a Biochemist,* Kornberg joyfully reminisced on the four-year period between 1948 and 1952. "I was at the laboratory bench all day with few interruptions," he wrote. "Each night, after our children were coaxed to bed I would sit in an easy chair with pencil and paper, and Sylvy might ask: 'What are you doing?' 'Thinking,' I'd answer. I would be designing experiments and preparative procedures for the next day, with fallback plans to cope with failures and disasters."[40] Kornberg maintained this habit for the rest of his professional life. Daily life in the laboratory was nothing short of blissful and he enjoyed and appreciated the efforts of

With Sylvy Kornberg at the University of Washington.

his research assistant Bill Pricer, who he referred to as "able, amiable, and devoted."[41] The only interruption to the day's work was a noon break for lunch during which Leon Heppel, Bernie Horecker, Herb Tabor (another distinguished scientist at the NIH) and he took turns presenting a scientific paper that they had agreed to read the previous night.

Herbert (Herb) Tabor is another illustrious figure in American biochemistry, who spent his entire research career at the NIH. Born in 1918 and hence approaching his ninth decade of life at the time of this writing, Tabor married Celia White in 1946. The two met on a streetcar in Boston when Herb was in medical school — and a lifelong collaboration began. Beginning in 1952 the Tabors worked together at the NIH studying polyamines. The couple became authorities on the microbial biosynthetic pathways of polyamines, the enzymes that catalyze steps in these pathways, and the functions of polyamines. They demonstrated that polyamines are multifunctional, being required for growth, sporulation, the maintenance of the killer double stranded RNA virus, protection against oxidative damage and elevated temperatures, the fidelity of protein biosynthesis, and the maintenance of mitochondria.[42]

Tabor's most lasting legacy will undoubtedly be his more than a half century of service to and leadership of the *Journal of Biological Chemistry* (JBC). He was a member of the JBC editorial board from 1961 to 1966 and an Associate Editor from 1967 to 1968. He became Acting Editor of the journal in 1969 and eventually Editor in 1971. Tabor sustained a philosophy that the journal should remain true to the goals of its founders: to publish papers in any area of biochemistry without regard to trendiness or cliquishness. He guided the journal through an enormous growth period from approximately 1,520 submissions in 1971 to over 15,000 in 2005. Importantly, Tabor led a technical revolution in science publishing with the 1995 debut of *JBC Online*, the first science journal to be published online.[43]

"Arthur and Sylvy and my wife Celia and I had a very close personal relationship" Tabor commented. "And at work Arthur and I discussed things a lot." Commenting on the lunch seminar group mentioned above, Tabor related:

> One Passover I could not bring food to our daily lunch seminar because we had a rather strict Kosher home. So Arthur and Sylvy made me sandwiches. When there were special occasions, like a birthday or an anniversary, Sylvy would make him a special lunch. I was once sitting next to Arthur and without realizing it I ate his lunch!

"There was very little by way of classical music in Washington in those days," Tabor continued.

> There was once a concert featuring an orchestra from Budapest that the lunch group was very keen to hear. But to get tickets one had to get down to the concert hall at about 6:00 am to stand in line. Having devoted time to that activity we decided to forgo the seminar lunch. So we had a seminar without lunch on G. Street!

An ardent admirer of his friend and colleague, Tabor related that

> Arthur was constantly on the go. He never sat down. So much so that when I took a break to sit down at my desk I often had the feeling;

Herbert (Herb) Tabor.

"Arthur would never do that." In those days he was solely interested in his work with no interest in anything else. When he went to Stanford in later years I was sure that his highest priority was to turn out a first rate biochemistry group that was not interfered with or diluted by others. I know some of the criticisms about this. But the truth be told Arthur did not want his department diluted by people that had a borderline interest in what was going on in his department and who may not concur with "family decisions." Arthur never did anything wrong or immoral. He never did things for political gain. And I never had a sense from all his writing that he undercut or downgraded people. Everything was very straightforward.[44]

References

1. Carl Ferdinand Cori. http://en.wikipedia.org/wiki/Carl_Ferdinand_Cori
2. Ibid.
3. Ibid.
4. Carl and Gerty Cori and Carbohydrate Metabolism, https://www.acs.org/content/acs/en/education/whatischemistry/landmarks/carbohydratemetabolism.html

5. Ibid.
6. Ibid.
7. Ibid.
8. Carl Ferdinand Cori. http://en.wikipedia.org/wiki/Carl_Ferdinand_Cori
9. Ibid.
10. Ibid.
11. AK Interview with Sally Smith Hughes, Program in the History of the Biosciences and Biotechnology, Biochemistry at Stanford, Biotechnology at DNAX, Arthur Kornberg, 1997, p. 19.
12. Carl and Gerty Cori and Carbohydrate Metabolism, https://www.acs.org/content/acs/en/education/whatischemistry/landmarks/carbohydratemetabolism.html
13. Friedberg EC. (2014) *A Biography of Paul Berg — The Recombinant DNA Controversy Revisited.* World Scientific Publishing Co., Singapore, p. 56.
14. Carl and Gerty Cori and Carbohydrate Metabolism, https://www.acs.org/content/acs/en/education/whatischemistry/landmarks/carbohydratemetabolism.html
15. AK Interview with Sally Smith Hughes, Program in the History of the Biosciences and Biotechnology, Biochemistry at Stanford, Biotechnology at DNAX, Arthur Kornberg, 1997, p. 12.
16. Kornberg A. (1991) *For the Love of Enzymes: The Odyssey of a Biochemist.* Harvard University Press, p. 65.
17. Ibid., p. 60.
18. Ibid., p. 69.
19. Ibid.
20. Ibid., p. 70.
21. Ibid., p. 71.
22. Ibid., p. 70.
23. Tom Kornberg, personal communication.
24. Exton JH. (2013) *Crucible of Science: The Story of the Cori Laboratory.* Oxford University Press, p. 115.
25. Ibid.
26. Oliver H. Lowry, http://en.wikipedia.org/wiki/Oliver_H._Lowry
27. Kresge N, Simoni RD, Hill RL. (2005) The Structure of NADH: the Work of Sidney P. Colowick. *J Biol Chem* **280**: e36. [http://www.jbc.org/content/280/39/e36]
28. Kornberg A. (1991) *For the Love of Enzymes: The Odyssey of a Biochemist.* Harvard University Press, pp. 72–73.

29. Ibid., p. 74.
30. Ibid., p. 75.
31. Letter from Arthur Kornberg to Carl Cori, Oct. 6, 1948. Reproduced with permission from the Department of Special Collections, Stanford University Libraries.
32. Kornberg A. (1991) *For the Love of Enzymes: The Odyssey of a Biochemist.* Harvard University Press, p. 82.
33. Ibid., pp. 82, 83.
34. Ibid., p. 82.
35. Glycolysis, http://en.wikipedia.org/wiki/Glycolysis
36. Pfizer Award in Enzyme Chemistry, http://en.wikipedia.org/wiki/Pfizer_Award_in_Enzyme_Chemistry
37. Letter from Herman Kalckar to Arthur Kornberg, Oct. 24, 1950. Reproduced with permission from the Department of Special Collections, Stanford University Libraries.
38. Kornberg A. (1991) *For the Love of Enzymes: The Odyssey of a Biochemist.* Harvard University Press, p. 82.
39. AK Interview with Sally Smith Hughes, Program in the History of the Biosciences and Biotechnology, Biochemistry at Stanford, Biotechnology at DNAX, Arthur Kornberg, 1997, p. 62.
40. Kornberg A. (1991) *For the Love of Enzymes: The Odyssey of a Biochemist.* Harvard University Press, p. 79.
41. Ibid.
42. Kresge N, Simoni RD, Hill RL. (2007) The Biosynthesis of Polyamines: The Work of Herbert Tabor. *J Biol Chem* **282**: e26–e28. [http://www.jbc.org/content/282/32/e26]
43. Ibid.
44. ECF Interview with Herb Tabor, September 2013.

CHAPTER FOUR

Washington University

In 1952 Kornberg was comfortably ensconced at the NIH. His research was progressing to his satisfaction and his name was becoming a household one in biochemical circles. He had no thoughts of relocating anywhere. But he was becoming aggravated that in his capacity as a section chief he was required to periodically interact with senior NIH administrators — who he referred to as "less-than-inspiring."[1] Additionally, the Institute was then building a Clinical Center dedicated to research on patients with specific diseases of interest, and Kornberg harbored concerns that basic biomedical investigation, then an overarching feature of life at the NIH, would undergo a transformation to more clinically oriented agendas. In terms of his stability at that institution it might be said in retrospect that Kornberg was then ripe for the plucking. Nonetheless, having never held a faculty position at an academic institution he was more than a little surprised to be invited by Carl Cori and Oliver Lowry at the Washington University School of Medicine in St. Louis to consider assuming the recently vacated chairmanship of the Department of Microbiology (formerly the Department of Bacteriology and Immunology).

"There were two elements that led me to accept their offer," Kornberg related in later years.

> I was becoming disillusioned with the less-than-inspiring administration of the NIH and I was displeased with the reality that I was at a level where I had to intersect with them. Secondly, the NIH was building its Clinical Center and I thought that this was ominous because the

basic science that was dominant at the NIH would now become very clinically and practically oriented.[2]

Kornberg expressed some of this ambivalence in a letter to Edwin Krebs penned in early 1949:

> The Institute is in a state of very rapid change, and it is difficult to predict what its strong points and weak ones will be a year from now, but I am rather hopeful that it will continue to be a very good place to work.[3]

Kornberg's eldest son Roger reinforced this sentiment:

> A pivotal situation that lay behind all this maneuvering was the reality that Arthur had been rocking the boat at the NIH for some while. He wasn't pleased with the existing structure and had been pushing for more obvious leadership authority there. Despite permission to organize Bernie Horecker, Leon Heppel and himself as an informal enzymology group he was always pushing for more and faster and his impatience raised the hackles of some in senior positions in the NIH administration. When they stalled he would push with "why?" "why?" "why?" He wanted things to happen today, not tomorrow. Eventually they told him: "If you want your own division you'll have it." But, what was unspoken was that "it's the last thing you're going to get from the NIH." I've always believed that part of the reason Arthur moved to St. Louis was that he'd run up against too many administrative barriers at the NIH for his liking.[4]

****

Washington University owned a highly distinguished reputation dating back to the mid-1800s. By 1952 the institution could boast six Nobel Laureates: Arthur Compton (of the Compton effect), Edward Doisy for his work on vitamin K, neurobiologists Joseph Erlanger and Herbert Spencer Gasser, and Carl and Gerty Cori. Subsequent Nobel Laureates associated with the university faculty at one time or another (besides Kornberg, see later) include Paul Berg, Stanley Cohen, Christian de Duve, Al Hershey, Luis Leloir, Rita Levi-Montalcini, Daniel Nathans, Severo Ochoa, George Snell, Hamilton Smith and Earl Sutherland.[5]

Silently debating the pros and cons of accepting a chairmanship and its multiple attendant demands was complicated by the usual concerns that accompany opportunities to move to new and more elevated professional positions. Aside from his mistaken assumption that the emergence of a Clinical Center at the NIH would stifle basic research there, another false conjecture emerged from Kornberg's conviction that the level and extent of the bureaucracy associated with university appointments would be more inspiring than that at the NIH, an opinion that rapidly dissolved when he settled into his new life as a department chair. Kornberg's then relatively naive understanding of academic politics failed to embrace the appreciation that universities in general do not differ much from non-academic organizations in terms of administrative issues. In referring to executive committee work at Washington University he once commented:

> As a member of the Executive Committee I looked forward to the opportunity to discuss science and education. Rather, there were debates about nurses' salaries and things like that. It was no more inspiring administratively that it had been at the NIH.[6] I often wondered in the succeeding years whether I had made a great blunder in moving from the NIH, where I was so comfortable, had such excellent facilities and things were going so well. The academic heaven in St. Louis never materialized. I became chairman of a department dealing with subjects in which I had no special knowledge. I was not a bacteriologist or an immunologist. In fact, after two or three years in St. Louis I wrote a memo to the dean of the medical school proposing that instead of having students who wanted to be sports medicine doctors and private practitioners we should appeal to them for a career in academic medicine and research medicine — without diminishing their capacity to practice medicine. My memo was ignored. I wanted to explain why biochemistry was essential for the teaching and practice of microbiology. But there was no textbook at that time that validated that approach; no book the students could consult to capture the new attitude about the importance of biochemistry and

genetics for learning about microbes. The students were very unhappy because ours was a novel, unorthodox and possibly wrong approach to medical microbiology. We had no textbook, no tradition, no precedent.[7]

Kornberg's disappointment with the state of affairs at Washington University was by no means ameliorated by the fact that when he arrived in St. Louis the promised number of appropriately renovated laboratories was not in evidence. Those in the Department of Microbiology were nothing to rave about — to say the least. They were 50 years old and seriously decrepit. Some broken windows had not even been repaired and had to be plugged with rags as a stopgap measure. As Kornberg described it,

> bare light bulbs hung from twenty-foot-high ceilings, electric outlets were sparse, the tiny sinks leaked and wooden bench tops were pitted and corrugated. Three laboratories were tardily refurbished and one more new laboratory was needed.[8]

In due course he irately informed the dean of the medical school that if renovation of the laboratories was not completed in a month, he would leave St. Louis! Carl Cori, who understood Kornberg's personality better than most at Washington University, independently warned the dean that Kornberg was absolutely serious about leaving if promises were not rapidly translated to action. Fortunately the NIH extramural research program was then getting off the ground and relatively lavish research funding was available. With the help of NIH grants Kornberg was ultimately able to build a well-equipped Department of Microbiology.

In the fall of 1953 Kornberg and his staff were confronted with the task of teaching microbiology to 120 medical students in their second year of study. The faculty remnants of the previous department, two or three people, taught old-fashioned bacteriology in which one carried out procedures to identify bacteria by staining — or some other outmoded means. The focus was on the pathogenesis of the disease for which the particular bacillus or coccus was responsible. "When I listened to a few of the lectures it was apparent how inappropriate it was to teach microbiology in these very narrow practical ways," Kornberg stated. "I was accustomed to the old orientation from having been a medical student, but that was fifteen years earlier and I was now imbued with biochemistry and genetics."[9]

Aware that he would be providing leadership to a microbiology department, Kornberg was determined to beef up his knowledge of this field. Among other concerted efforts he attended the revered microbiology course at the Stanford Oceanographic Institute in Pacific Grove, California, taught by the esteemed microbiologist Cornelius Bernardus (C.B.) van Niel.[10],* "Van Niel was an extraordinary figure and a teacher of the kind that one never finds anymore," Kornberg related. "He could lecture for eight hours a day and enthrall you with accounts of heroes and villains in science; the Dutch microbiologists were always his heroes."[11] "He dwelled on the 'good' microbes in the environment and forbade the mention of pathogens, except those few that figured prominently in the history of microbiology."[12]

Aside from the bliss of exchanging a summer in St. Louis for a location on the shores of the Pacific Ocean, van Niel's course provided an outstanding historical overview of microbiology and a powerful antidote to medically oriented bacteriology. That summer was additionally pleasurably augmented by the presence of Kornberg's wife Sylvy and their sons in Pacific Grove, who of course were equally delighted to trade a summer in St. Louis for the balmy shores of the Pacific Ocean. During the course Kornberg presented a seminar on the isolation of enzymes from the cellular juices of microbes. "This is beautiful work," van Niel told him. "I know it needs to be done. I myself would not have the heart to grind up the little beasties!"[13]

---

*The Dutch microbiologist Cornelius Bernardus (C.B.) van Niel arrived at the Hopkins Marine Station of Stanford University located on the Monterey Peninsula, California in 1930. van Niel developed a revolutionary concept of the chemistry of photosynthesis that was to influence research on the topic for many years. He also was the first in the United States to teach general microbiology, an effort that resulted in the great blossoming of microbiology in the 20th century. Many distinguished microbiologists were influenced by him directly or indirectly. [http://highered.mcgraw-hill.com/sites/dl/free/0072320419/20534/vanniel.html].

"After a year or two the students who had taken biochemistry in their first year dubbed our course 'Biochemistry II'. It was not well received," Kornberg related.[14] "A fifth column of several people left over from the old department would tell the students that they weren't learning medical bacteriology and were being deprived of exposure to information about syphilis and other diseases that would be crucial to becoming legitimate MDs.[15] But in fact in later years both practicing MDs and those in academic medicine would refer to that course as the most memorable and enlightening course that they had had at Washington University. Indeed, in late December 1980, more than two decades after he left St. Louis, Kornberg received a letter from James Creveling, a former medical student at Washington University who wrote:

Dear Dr. Kornberg,

Since my graduation from Washington University School of Medicine in 1959, I have been more aware with each passing year of the far-sightedness that men like yourself showed in teaching us the basics. At the time, we thought 'What has this biochemistry got to do with microbiology.' --------------------- I thank you for your part in shaping my life.[16]

"This sort of opinion from some of the students was not only due to the curriculum, but also to the spirit of the young people who were teaching the course, who were imparting subjects that they hadn't known anything about, but were learning it with novel insights," Kornberg related.[17]

****

The first investigators to join Kornberg's department in St. Louis were postdoctoral fellows Osamu Hayaishi, who had joined Kornberg's laboratory at the NIH in 1950 and who Kornberg promoted to the rank of Assistant Professor, and Irving Lieberman, also with Kornberg at the NIH, who was given the title of Instructor. Kornberg had great admiration and respect for Hayaishi. In 2006 he wrote a glowing tribute to the Japanese investigator entitled "Osamu Hayaishi: Pioneer of the Oxygenases, the Molecular Basis of Sleep and Throughout a Great Statesman of Science." "One of the happiest parts of my scientific life has been the friendship and inspiration of Osamu Hayaishi," Kornberg wrote.

> Typical of his generous nature, he has referred to me as his mentor, but a strong case will be made that in the balance of credits for a tutorial role, it was actually the reverse. Osamu came to my laboratory at the NIH in 1950 ostensibly to learn enzymology. While doing that he taught me the power of microbiology. He introduced me to the technique of enrichment culture of soil bacteria as a means to discover novel enzymes and pathways, in what has revealed over and over again the universality of biochemistry. Among the lessons I have learned in my lifetime of science, I place this focus on microorganisms, second only to enzymology for gaining insights into biological events in nature.[18]

In his later years back in Japan, Hayaishi challenged the prevailing view that the oxidation of a substance by the action of dehydrogenases was the ultimate source of oxygen. He demonstrated the existence of a class of oxygenases and their ubiquity in performing a variety of functions throughout nature, from microbes to humans.[19] He also carried out pioneering work on the molecular and genetic basis of consciousness. "His discoveries of specific prostaglandins for wakefulness and for sleep have biochemical, genetic and physiological foundations unmatched in neuroscience," Kornberg commented.[20]

In the next three years, seven people joined the faculty of the Department of Microbiology at Washington University. Melvin Cohn, David Hogness and Dale Kaiser arrived following postdoctoral training with Jacques Monod and Francois Jacob at the Pasteur Institute in Paris. Robert (Bob) deMars came from Salvador Luria's group at the University of Illinois and Jerard Hurwitz came from the NIH. Paul Berg and Robert (Bob) Lehman became faculty members after completing postdoctoral fellowships with Kornberg.[21] Berg, Lehman, Kaiser, and Hogness formed enduring relationships with Kornberg. As recounted in a later chapter all four of these individuals joined the faculty of his biochemistry department at Stanford University in 1959.

Paul Berg had just completed a postdoctoral stint with Herman Kalckar in Copenhagen and was sufficiently enamored with the opportunity of working with Kornberg that he overcame his utter disdain for St. Louis, especially its summers. Berg, who had recently obtained his PhD from the Department of Biochemistry at Western Reserve University in Cleveland Ohio, had rebuffed chairman Harland Wood's insistent advice that he pursue postdoctoral work with the Coris at Washington University because of his aversion to living in St. Louis. But, adamant as he was in joining Kornberg's laboratory following his stint in Denmark, he moved to that mid-western city — albeit with some reluctance.

"In anticipation of finishing up in Cleveland I wrote to Arthur sometime in early 1952 asking if I could join his lab at the NIH starting in the Fall of 1953, as I already had acceptance from Kalckar for me to go to Copenhagen in the Fall of 1952," Berg related.[22]

> I had seen Arthur's picture on the cover of *Chemical and Engineering News* after he won the Paul Lewis Award. In the article I learned that like me he had graduated from Abraham Lincoln High School. When he came to Cleveland in early 1952 to give a seminar I introduced myself to him as another graduate of that high school. His work and charm impressed me so I decided to apply for a position as a postdoctoral fellow in his NIH laboratory. He accepted me and I went off to Copenhagen with assurance that I would join him at the NIH when I returned. A letter informing me that he would be moving to St. Louis in Jan. 1953 arrived while I was still in Copenhagen.[23]

Over the years Paul Berg and his wife Millie formed an enduring relationship with Arthur and Sylvy Kornberg. In a memorial article by Berg not too long after Kornberg's death in 2007, Berg wrote:

> My relationship with Arthur Kornberg went beyond that of a student and mentor. Soon after I joined his lab as a postdoc, my wife and

I were embraced as members of his family and we shared many of the special occasions and achievements they celebrated. ------------------------------- Over the course of time, close friendship took the place of mentorship and discussions about art, music, and politics filled in times when our discussions about experiments lagged.[24]

Berg's strong background in biochemistry notwithstanding, he gallantly pitched in with teaching microbiology at Washington University — as did Kornberg himself. Kornberg taught about bacteriophages and later admitted that "teaching that course proved to be instrumental in my later discovery of DNA polymerase. I became curious about the growth of phages and how they replicated their DNA."[25]

Robert (Bob) Lehman, a veteran of World War II, saw front-line combat action in Europe and was awarded a Purple Heart in recognition of a wound sustained during the allied invasion of Germany from France. "I was a sergeant in the 3rd Infantry Division which was part of the 7th Army that participated in the invasion of Southern France in August 1944," Lehman related.

> We fought our way up the Rhone Valley through the Voges Mountains during the winter of 1944–45 and liberated the Alsation cities of Colmar and Strasbourg. We then crossed the Rhine and fought our way to the Austrian Border when the war ended in May 1945. The 3rd Division suffered the most casualties, had the most days of combat and received the most decorations of any Division in the US Army. Before I joined the 3rd Division it had fought in North Africa, the invasion of Sicily and the invasion of Italy, first at Salerno then at Anzio. Miraculously, I survived with only minor wounds.

The close fighting against the retreating but defiant German Army in the dense pine forests of the Voges Mountains was both nerve-wracking — and lethal. Lehman's infantry company, about 200 strong in August 1944, was reduced to about 30 soldiers by May 1945.[26] At the war's end elements of the 7th Infantry Regiment serving under the 3rd Infantry Division captured Hitler's retreat home *The Berghof* at Berchtesgaden.[27]

Robert (Bob) Lehman.

Upon returning to the United States, Lehman availed himself of the benefits of the GI bill and attended college and then graduate school at Johns Hopkins University in Baltimore, Maryland. While attending a meeting in Atlantic City Lehman encountered Irving Lieberman, who, as mentioned above, had first joined Kornberg's laboratory while he was at the NIH and had followed him to St. Louis. Lehman was intrigued to hear about the research in Kornberg's laboratory and "decided on the spot that that was the sort of work I wanted to do. I wrote to Arthur, who offered me a postdoctoral position."[28]

When Kornberg contacted Andre Lwoff at the Pasteur Institute to determine whether there was anyone in his research group who might be able and interested in teaching virology to medical students, Lwoff sang the praises of Dale Kaiser. "Bob DeMars, who was teaching virology in Arthur's microbiology course at Washington University stopped teaching for a while. I think he was having a dilemma about what he wanted to do — without quitting his position," Kaiser related.

While Demars was away Arthur contacted Lwoff to determine whether there was anyone else in his group at the Pasteur who might be able to take over teaching virology. So he appointed me as an Acting Instructor. I didn't have any specific plans then. I was simply looking for opportunities that might present themselves. The only way I could think easily about science was by dealing with genetics. Arthur was working on DNA replication and that seemed like a great opportunity to work on the chemistry behind the genetics of DNA replication. So his offer was very attractive to me.[29]

"Arthur was something of a taskmaster — but a fair one," Kaiser commented. "In the early days at Washington University there was a departmental inspection every Saturday morning. Arthur would go around the department and would feel the tops of cabinets for dust. And if he found one (or more) that were outside your lab you'd be told to clean it up!"

The burden of teaching microbiology and virology and coping with the machinations of the medical school administration notwithstanding, Kornberg enjoyed the scientific milieu for which Washington University was justifiably renowned. Besides the Coris, Joseph Erlanger and Herbert Gasser (who adapted the cathode-ray oscillograph for the study of nerve action potentials that led to work for which the pair received the Nobel Prize in Physiology or Medicine in 1944)[30] were among many distinguished members of the basic biomedical community who Kornberg relished interactions with.

Outside of sporadic bouts of aggravation or annoyance such as those just mentioned in relation to the delayed renovation of his St. Louis laboratories, Kornberg, who, on occasion could be unmistakably tough and obstinate, was generally polite, a personality feature likely inculcated by his parents during his early upbringing. His courtesy certainly leaps from the pages of his voluminous correspondence. Kornberg routinely offered time and energy to the composition of notes of thanks or appreciation to those one might least suspect, frequently in his own hand. In late 1948 for example, after acquiring one of many shipments of yeast from the Research Division of the St. Louis-based company Anheuser-Busch Inc., Kornberg jotted a note to an individual of no special importance at the company expressing his appreciation for the attention his laboratory routinely received.

> "As many times before we are amazed and gratified by the extreme cooperativeness and efficiency of your organization," he wrote. "I mention this to emphasize our gratitude in having this large and excellent quantity of yeast that you prepared for us. I am terribly sorry that under Government regulations it would be impossible to cover the additional costs that you sustained in preparing this yeast."[31]

One doesn't often encounter that level of civility nowadays.

****

"My ambitions soared in 1950," Kornberg wrote in his autobiography.

> I had earlier found an enzyme in extracts of potato that inserts a water molecule into a respiratory coenzyme, cleaving it into two nucleotide components. By using this apparently ordinary enzyme to produce a novel starting material I could discover how coenzymes are made in cells and with it a basic, biochemical theme used in the biosynthesis of lipids, steroids, proteins and nucleic acids. Having learned how a nucleotide, the building block of RNA and DNA is built into a coenzyme, I was led to wonder if many thousands of them might be assembled to make the chains of the nucleic acids.[32]

Such was the awakening of Arthur Kornberg's interest in nucleic acids, specifically DNA. Much of the rest of this biography details Kornberg's heroic achievements in nucleic acid biochemistry, studies that systematically examined the enzymatic reactions by which nucleotides are synthesized and converted to triphosphates, his discovery of the first DNA polymerase, and his identification and characterization of a large library of other proteins required for DNA replication in prokaryotes, studies initiated at the NIH, continued in St. Louis and brought to full fruition at Stanford University.

At Washington University. *Front row*: Kornberg is 3rd from right. Gerard (Gerry) Hurwitz is seated to the left of Kornberg; David Hogness is at the extreme right. *Middle row*: Bob Lehman is on the extreme left, Dale Kaiser is 3rd from the left and Paul Berg is 4th from the right.

References

1. AK Interview with Sally Smith Hughes, Program in the History of the Biosciences and Biotechnology, Biochemistry at Stanford, Biotechnology at DNAX, Arthur Kornberg, 1997, p. 12.
2. Ibid.
3. Letter to Edwin Krebs, Feb. 17, 1949. Reproduced with permission from the Department of Special Collections, Stanford University Libraries.
4. ECF Interview with Roger Kornberg, May, 2014.
5. History and Traditions — Washington University in St. Louis, http://wustl.edu/about/history
6. AK Interview with Sally Smith Hughes, Program in the History of the Biosciences and Biotechnology, Biochemistry at Stanford, Biotechnology at DNAX, Arthur Kornberg, 1997, pp. 13, 15, 17.

7. Ibid.
8. Kornberg A. (1991) *For the Love of Enzymes: The Odyssey of a Biochemist.* Harvard University Press, p. 179.
9. AK Interview with Sally Smith Hughes, Program in the History of the Biosciences and Biotechnology, Biochemistry at Stanford, Biotechnology at DNAX, Arthur Kornberg, 1997, p. 14.
10. Kornberg A. (1991) *For the Love of Enzymes: The Odyssey of a Biochemist.* Harvard University Press, p. 105.
11. AK Interview with Sally Smith Hughes, Program in the History of the Biosciences and Biotechnology, Biochemistry at Stanford, Biotechnology at DNAX, Arthur Kornberg, 1997, p. 13.
12. Kornberg A. (1991) *For the Love of Enzymes: The Odyssey of a Biochemist.* Harvard University Press, p. 105.
13. Ibid., p. 106.
14. AK Interview with Sally Smith Hughes, Program in the History of the Biosciences and Biotechnology, Biochemistry at Stanford, Biotechnology at DNAX, Arthur Kornberg, 1997, p. 14.
15. Ibid.
16. Letter from James Creveliny to Arthur Kornberg, Dec. 22, 1980. Reproduced with permission from the Department of Special Collections, Stanford University Libraries.
17. AK Interview with Sally Smith Hughes, Program in the History of the Biosciences and Biotechnology, Biochemistry at Stanford, Biotechnology at DNAX, Arthur Kornberg, 1997, p. 14.
18. Kornberg A. (2006) Osaumu Hayaishi: Pioneer First of the Oxygenases, then the Molecular Basis of Sleep and Throughout a Great Statesman of Science. *Life* **58**: 253.
19. Ibid.
20. Ibid.
21. Kornberg A. (1991) *For the Love of Enzymes: The Odyssey of a Biochemist. Harvard University Press*, p. 179.
22. ECF Interview with Paul Berg, Nov. 2013.
23. Berg P, personal communication, Dec. 2014.
24. Berg P. (2009) Arthur Kornberg. *Proc Amer Phil Soc* **153**: 468–475.
25. AK Interview with Sally Smith Hughes, Program in the History of the Biosciences and Biotechnology, Biochemistry at Stanford, Biotechnology at DNAX, Arthur Kornberg, 1997, p. 15.

26. Robert Lehman, personal communication.
27. 3rd Infantry Division (United States), http://en.wikipedia.org/wiki/3rd_Infantry_Division_(United_States)
28. ECF Interview with Robert Lehman, Oct. 2014.
29. ECF Interview with Dale Kaiser, Aug. 2013.
30. The Nobel Prize in Physiology or Medicine, 1944. http://www.nobelprize.org/nobel_prizes/medicine/laureates/1944/erlanger-bio.html
31. Letter to Kenneth Holt, Anheuser-Busch, Dec. 23, 1948. Reproduced with permission from the Department of Special Collections, Stanford University Libraries.
32. Kornberg A. (1991) *For the Love of Enzymes: The Odyssey of a Biochemist.* Harvard University Press, p. 122.

CHAPTER FIVE

The Lure of Nucleic Acid Enzymology

As the middle of the 20th century approached, no one had the faintest notion of how cells make DNA or RNA. Nor was the biosynthesis of the nucleotide building blocks of these macromolecules even established. "I freely admit that when the Watson–Crick paper came out I was not electrified by it, as I should have been," Kornberg stated.

> But I was ambitious enough to want to know more about the enzymology of nucleic acids from the bottom up; how each building block is made and then assembled into DNA and RNA. Actually, I approached the problem in a roundabout way. The phosphodiester linkage found in RNA and DNA is also found in another class of compounds, the phospholipids. So I studied phospholipids because I thought that if I could understand how that linkage is made in a simpler and more accessible molecule I might understand how it is made in the chains of DNA and RNA.[1]

Questions about the biosynthesis of nucleic acids churned in Kornberg's thoughts. "Was the backbone assembled first and were the bases attached later?" he asked rhetorically.

> Was each link added to the chain as a single nucleotide? If so, was the phosphate in each component nucleotide initially attached to carbon number 3 or 5, or to either one randomly or in a cyclic form to both? But first we had to know the building blocks of the nucleic acids. It was

not at all obvious in 1950 what they may be. In as much as others were already pursuing purine biosynthesis, I decided to go after pyrimidines.[2]

Kornberg first explored the synthesis or degradation of pyrimidines in extracts of liver and yeast — with little success. So he resorted to the use of enrichment cultures with the assistance of Osamu Hayaishi who, as mentioned in the previous chapter, had recently joined Kornberg's NIH laboratory as a postdoctoral fellow, and who was an authority on enrichment culture. The development of enrichment culture technology in the very early years of the 20th century is credited to the Dutch microbiologist and botanist Martinus Beijerinck. The logic of the procedure is straightforward. One incubates soils (or other sources likely to harbor bacteria) in a liquid medium supplemented with a compound of particular interest and hopes to observe the emergence of strains that can survive under such conditions. A Dutch contemporary of Beijerinck, Lourens Baas Becking, succinctly summarized the utility of enrichment culture with the aphorism "everything is everywhere but the environment selects."[3]

Hayaishi dutifully demonstrated the power of enrichment culture to Kornberg by scraping some soil from a car tire and suspending it in flasks containing a salt solution, each with a different precursor. The following

Osamu Hayaishi.

morning Kornberg was astonished to find that the precursor had vanished from the solution, which was now cloudy with swarms of bacteria.[4] Some of Hayaishi's enrichment cultures supplemented with cytosine, uracil or thymine pleasingly survived under these conditions. But to Kornberg's disappointment the pyrimidines were converted to nothing more interesting than barbiturates.

In 1951, two years prior to his move to St. Louis from the NIH, Kornberg began exploring the biosynthesis of orotic acid, a compound that structurally resembles uracil and was then considered a likely precursor of that pyrimidine. He made little progress with extracts of animal or yeast cells. Nor was he able to obtain an enrichment culture in the summer of 1951. At this point he sought help from Horace Albert (Nook) Barker at the University of California at Berkeley. A renowned microbiologist who had trained under the legendary C. B van Niel, Barker made his mark in the early 1950s by elucidating the biochemical function of vitamin B12. In 1944 he was a member of a team that discovered the role of enzymes in the synthesis of sucrose by deploying one of the earliest laboratory uses of radioactive carbon-14 tracers — a technique that Barker helped pioneer. In 1968, he received the National Medal of Science from President Lyndon Johnson for his many scientific contributions, particularly those involving vitamin B12.[5] Influenced by van Niel, Barker also made outstanding contributions to microbiology — including the use of enrichment cultures.

Barker took Kornberg to his favorite brackish pond near the shores of the San Francisco Bay where he scooped up a few spoonfuls of mud and established an enrichment culture from which he isolated a novel microbe adept at converting orotic acid to smaller units that later turned out to be valid precursors in the synthesis of nucleic acids — but not of uracil. Kornberg later resorted to the use of liver and yeast extracts in his pursuit of the biosynthesis of uracil. Barker named the new strain *Zymobacterium oroticum*, prompting Kornberg to quip that Barker's adoption of that name instead of *Zymobacterium kornbergii* dispensed with his best chance for immortality![6]

As also mentioned in the previous chapter, Kornberg had additionally been joined in his NIH laboratory by an eager postdoctoral fellow, Irving

Arthur Kornberg at the blackboard.

Lieberman (affectionately referred to as Lieb). Young Lieberman made an indelible impression on Kornberg at their very first meeting. "On the day he arrived at my laboratory in Bethesda to start his postdoctoral fellowship it was already near quitting time," Kornberg related.

> He had driven across the country with his wife and infant son and come directly to the laboratory. With barely a greeting, he wanted to know about research projects he might work on, and he instantly chose orotic acid. I guessed it would take a week or so to get settled — and offered to help. Lieb interrupted, "Can we go over an experimental plan now?" I was already late for dinner at home. "Sure," I said, "the first thing tomorrow." "Why tomorrow, Doc? We can get some cultures started tonight."[7]

Kornberg could not dissuade Lieberman from sterilizing media and starting a series of culture flasks right then and there. Later he learned that his wife and child had been waiting for him in their baggage-stuffed car those several hours in a darkening, abandoned parking lot.

Kornberg and Lieberman struggled to obtain respectable yields of uracil from orotic acid — until they decided to mix yeast and liver extracts that had been individually used as starting materials, to address the possibility that any factor(s) missing in one extract would be supplied by the other. "Wow!" Kornberg wrote. "The reaction was explosive — hundreds of times greater than before."[8] As it turned out, one enzyme was abundant in the yeast extracts but failed to survive in the liver extracts. When mixed, the liver and yeast extracts complemented each other. Further studies revealed that a liver enzyme used ATP to produce an activated form of ribose phosphate, which the yeast enzyme deployed to release carbon dioxide from orotic acid to produce uracil.[9]

Lieberman subsequently found enzymes in extracts of liver and yeast that converted uridine monophosphate to the di- and then the triphosphate forms. Further studies yielded the biosynthetic pathways for cytidine triphosphate. Kornberg and Lieberman with the able assistance of Ernie Simms, Kornberg's devoted and capable research assistant, also found an enzyme that condensed phosphoribosyl pyrophosphate (PRPP) with adenine to form adenine ribose phosphate, the adenine nucleotide, and yet another enzyme that did the same with guanine to form the other purine nucleotide. "Now it seemed we had the four building blocks of RNA and could press on to find the enzymes that assembled them into the giant molecule," Kornberg wrote.[10]

****

Kornberg's surprise about Lieberman's haste to don a white coat and get cracking in the laboratory immediately merits an interruption in the description of his pursuit of nucleic acid biochemistry to highlight a prominent feature of Kornberg's own compulsive nature about the use (or misuse) of time in the laboratory. Those who worked with Kornberg through the many years of his professional life beginning with his early days at the NIH well knew and often joked about his intolerance of what he called "wasting time," meaning any time away from the laboratory bench. He informed an interviewer that his obsession about time, especially what he considered as wasted time, originated during the postdoctoral year he spent with Severo Ochoa — where every day

The Kornberg family. From left to right are Roger Kornberg, Ken Kornberg, Sylvy, Arthur and Tom Kornberg.

mattered! "Ever since, I've told my students and people working with me, 'Hey, you wasted an afternoon; it will never come back.' I have always been preoccupied with time."[11]

This sentiment was strongly reinforced by the distinctive nature of biochemical research in contrast to research in other settings, such as hospital wards. "I could get a fact, or fail to find a fact in a matter of minutes or hours, rather than waiting for months, as in research with rats — or in humans when trying to sort out a very muddled set of circumstances presented by a patient," Kornberg related.[12] He was sometimes the subject of pointed teasing about his obsession with time, of which he was well aware. "People tell stories about me," he once related.

> Stories circulate that are largely embroidered or fictitious. One that has been told and is in essence correct, is about my obsession with time. I'm very conscious of the passage of time and certainly about the loss of time. That goes way back to when I was a teenager. Once, when my technician told me that he'd lost a sample but fortunately still had some of it and could do the experiment again, I might have commented: "But you did lose much of an afternoon with that accident!"[13]

"I remember meeting two senior scientists, both accomplished biochemists, and discussing a procedure they had suggested," Kornberg continued.

> I said: "I tried that and I wasted a whole afternoon on it." One turned to the other, and sarcastically quipped: "He wasted a *whole* afternoon on that procedure!" But it's true to this day. My clock is always in front of me. I may pick up an issue of *Nature* and tell myself: "I must not spend more than ten minutes reading this issue." Then I'm upset because there are so many things in the journal that are interesting and diverting. Yes, time has always mattered.[14]

Fully cognizant of the fact that many were well aware of his preoccupation with time, Kornberg nevertheless retained a sense of humor about this foible. Jerry Hurwitz relates that once when Kornberg was out of town, members of his Washington laboratory felt a compelling need to inform him of the loss of a preparation of radiolabeled deoxynucleotides that had involved much effort, expense — and time. Everyone knew that Kornberg would not be at all pleased about this catastrophe, so no one was especially keen to be the bearer of the bad news. Coin tosses eventually placed the responsibility on the shoulders of Kornberg's research assistant Ernie Simms. Simms made the call while the rest of the group sat around with bated breath. When the brief phone call concluded the group immediately asked Simms: "What did he say?" "Arthur said that he could understand the loss of the preparation — but he couldn't understand why we were wasting precious time calling him about it," Simms replied.[15]

Though time was indeed something of an obsession with Kornberg, his eldest son Roger, a keen observer of his eminent father's quirks and habits, never considered him a workaholic. "On the contrary," Roger related, "he never worked past 11PM, took regular hours for lunch and dinner — at least 2 hours when dining with his family — and devoted considerable blocks of time to leisure." Indeed, Kornberg loved to play tennis and when at Stanford he spent most Sundays at poolside in his nearby Portola Valley home, frequently entertaining the families of members of his biochemistry department, which he rotated so as to accommodate each on multiple occasions.[16] Nor was he loath to indulge his love of sports in general. When in St. Louis he regularly took his sons to St. Louis Cardinals baseball games. He maintained this tradition in California where he painlessly switched allegiance to the San Francisco Giants, managing to secure premier seats at Candlestick Park. In December 1987 he sent a personal note to Bob Luria, then the owner of the San Francisco Giants, pleading "for your help in getting a better season ticket location. [My sons and I] would very much appreciate a lower box location on either the first or after the third base lines, he wrote.[17] (However, Tom Kornberg points out that "we never had season tickets and never attended more than 2–3 games per year.") But there is agreement that Kornberg dutifully attended Stanford football games, for which the family had season tickets for over 30 years, and devoted time to watching sports on television. A letter to his son Tom states: "today is a big TV day; 10:00 am 49ers versus Eagles; 1:00 pm Giants vs. Pirates; 4:00 pm Laver vs Rosewall." "My father's work habits can be best summed up in a word — moderation," Roger Kornberg opined. "What was notable about his work habits were his self-discipline and concentration. When he worked he did so without distraction. He used every minute as efficiently as possible."[18]

Kornberg also spent extended periods of time away from his Washington University laboratory — if for no other reason than to escape the insufferable St. Louis summers.[19] These escapes began in the summer of 1956 when the entire Kornberg family together with Paul Berg and his own family spent a summer in British Columbia, though the trip did include time spent in Gobind Khorana's laboratory at the University of British Columbia.

> I had a reputation for working very hard, for being unforgiving in wasting an hour or an afternoon, and yet here I was in 1956, at the height of all the excitement about DNA polymerase (the discovery of which is recounted in the ensuing chapter) going off for two or three months to British Columbia with my family — taking off the whole summer. Several people asked, "Gee, how could you do that?" It was unthinkable for anybody, let alone me, to do that. Yet we had gone off repeatedly in summers, escaping the St. Louis heat — taking the family.[20]

Kornberg's impatience with other peoples' use of time, most notably with graduate students and postdoctoral fellows in his own laboratory, could sometimes escalate to intolerance bordering on frank anger. A brilliant graduate student Randy Schekman, who shared the Nobel Prize in Physiology or Medicine in 2013, said:

> I remember that a group of us were standing outside Arthur's office one Friday afternoon when I related my experimental results to him. He immediately asked when I was going to do the next obvious experiment, to which I responded that I'd probably get to it on Monday or Tuesday of the following week. "What's wrong with the week-end?" he asked. When I told him that Bill Wickner, I, and our wives were going skiing over the weekend he looked at me with abject scorn and said: "What am I running around here — some sort of fucking country club?"[21]

The encounter left young Schekman more than a little shocked.

"I remember that Randy had a bit of a stormy relationship with Arthur," Bob Lehman commented. "He would sometimes come to me complaining about him. But when Arthur died Randy wrote a glowing tribute in which he stated that as a graduate student there were times he didn't really appreciate what Arthur was doing for him. When all is said and done I believe that Randy really admired Arthur."[22] Indeed, shortly after Kornberg's death, Schekman published an obituary in which he wrote:

> Arthur tolerated me (barely) as an insecure and arrogant rookie graduate student. He taught me rigor and discipline and showed by example the power of biochemistry. I regret not telling him this, but I am not ashamed to say that I have spent the last 30+ years of my career trying to live up to his standard.[23]

Roger Kornberg relates a similar anecdote about Kornberg's intolerance about time away from the laboratory:

> On his 70th birthday celebration, when I expected that people would be especially nostalgic, many of them commented about how tough

Arthur had been with them. Maurice Bessman, who was a postdoctoral fellow with him working on DNA polymerase remarked that my father once asked him whether he was going to do such and such tomorrow and Maurice blithely responded that tomorrow was Saturday and he was going to do something with his family. My father gave him a very dark look and said: "You mean you're going out with your family on a Saturday?" So he could indeed be hard on people.

"I remember my mother commenting that Arthur was sometimes too hard on Tom, Ken and me as children, and that he should go easier on us otherwise 'your children will grow up disliking you,'" Roger Kornberg related. "'If you treat them like you do your postdocs, it will come back to haunt you,' she said."[24]

Some who knew Kornberg over extended periods refer to him as a complex individual. While his impatience and sometimes frank intolerance with the priority (or lack of it) for time spent in the laboratory could surface an abrasive attitude, he could be extremely solicitous and caring when it came to people's personal lives, and as just mentioned he was a most generous host in his private home. "He had a

An iconic photo of Arthur Kornberg.

remarkable sort of congeniality and genuine interest in people," Roger Kornberg related.

> Others used to remark about that. He would talk to their children and he'd take a serious interest in what they told him. That was something that was most noted. I remember someone telling me about a Gordon Conference where my father and others were standing in line to get dinner. Arthur noticed someone he knew with his obviously pregnant wife near the back of the line and immediately signaled them to change places with him. That sort of thing.[25]

Kornberg's generosity and his sensitivity to the needs and plight of others is borne out by many of his former graduate students and postdoctoral fellows, especially those to whom home was far removed from the Bay Area, but who received invitations to Kornberg's home around times of major holidays such as Thanksgiving and Christmas.

Following his death in 2007, Kornberg was the subject of numerous tributes and obituaries. A particularly touching and memorable gesture organized by the Stanford Medical School emerged as a booklet entitled *Tribute to Arthur Kornberg, MD*. The volume, arguably the center piece of a day-long program of oral, visual and musical reflections and remembrances that transpired on the Stanford campus on January 28, 2008, embraces close to 100 written tributes from former graduate students, postdoctoral fellows, faculty colleagues, relatives (including grandchildren) and friends. Teeming with poignant anecdotes, the volume is a memorable testament to the admiration, esteem and affection with which Kornberg was viewed during his life.

As for his sons, Roger Kornberg commented: "I don't remember him ever being what one would call hard. All my memories of him are of him smiling. Not only were we never spanked, but I don't even recall him being demanding."[26] "My brothers and I always considered Arthur our best friend," Roger confided.

> We could talk to him about anything that might be worrying us. In fact, he had a remarkable ability to anticipate if anything was "wrong" with any of us. He would somehow always sense if there was something

In the laboratory.

troublesome going on and he would bring it up before we ever mentioned anything. He was exceptionally sensitive to our moods. He was also of the firm belief that when parents intervene in their childrens' lives they can cause more bad than good. So he inculcated a firm sense of independence in all three of us at a very early age and encouraged us to have our own contrary opinions, a mentality appreciated by each of us.[27]

"My father was thoughtful about us to the extent that he deliberately discouraged any sort of hero worship or reverence that we might develop about him," Roger continued.

He was very sensitive to the risk of children being diminished by their parents' success. I remember countless times that when distinguished scientists came to visit, people like Feodor Lynen or Severo Ochoa, he would tell us that these people were very great scientists. We would naturally ask him: "Are *you* a great scientist?" His response was invariably that he was just an ordinary scientist. So much so that when he won the Nobel Prize, I seriously thought there must have been a mistake. It wasn't until I was quite far advanced in college that I began to realize that my father was indeed a great scientist. By then there was no danger of doing me any psychological damage because I was my own person with a healthy dose of arrogance and of disrespect for authority — all the things you need in your own right! Shortly after he won the Nobel Prize a reporter asked my father whether any of his children manifested scientific aptitude. He responded that none of us was unusual in that regard. In later years, he told me that he made that remark because he didn't want the person to think that we were 'freaks' of any sort, but that he didn't literally mean what he had said. That was the only time that I heard from him directly that he thought I did indeed have an aptitude for science.[28]

****

Kornberg began serious forays into the biosynthesis of nucleic acids by initially focusing on RNA. He explored the synthesis of RNA with an Israeli postdoctoral fellow Uri Littauer from the Weizmann Institute in Israel. The pair prepared ATP radiolabeled in the adenine moiety. "Uri observed that when the labeled ATP was incubated with an extract in *E. coli,* a small but significant amount of its radioactivity was rendered precipitable upon acidification,"[29] a hallmark of nucleic acids. Kornberg and Littauer inferred that they had observed a genuine enzymatic reaction.

> For one thing the activity in cell extracts was unstable. Additionally, the extent of the reaction was time-dependent — and finally the optimal temperature was in the range 30–37°C. We felt we were on to something very important — the enzymatic synthesis of RNA.[30]

In the spring of 1955, Kornberg was visited by Herman Kalckar, who brought "startling and unsettling news. For Littaeur it was devastating."[31] Severo Ochoa and his colleague Mariannne Grunberg-Monago had discovered an enzyme from the nitrogen-fixing bacterium *Azobactor vinelandii* that made RNA-like chains by condensing molecules of ADP or other nucleoside diphosphates. On the strength of this information, Kornberg and Littauer switched to using ADP instead of ATP in their studies with *E. coli*. "We were keenly disappointed," Kornberg related.

> We merely confirmed the existence of what Ochoa termed polynucleotide phosphorylase. We soon realized that we had been diverted from the

With Jack Griffith's pet tiger, named Arthur.

far more important discovery of the RNA polymerase that depends on ATP rather than ADP. By switching to ADP we missed the key enzyme of cell growth and function, the enzyme responsible for transcribing genes on the route to synthesizing proteins.[32]

References

1. AK Interview with Sally Smith Hughes, Program in the History of the Biosciences and Biotechnology, Biochemistry at Stanford, Biotechnology at DNAX, Arthur Kornberg, 1997, p. 18.
2. Kornberg A. (1989) Never a Dull Enzyme. *Ann Rev Biochem* 58: 1–31.
3. Lourens Baas Becking, http://en.wikipedia.org/wiki/Lourens_Baas_Becking
4. Kornberg A. (1991) *For the Love of Enzymes: The Odyssey of a Biochemist.* Harvard University Press, p. 101.
5. Saxon W. (2010) Horace Barker, 93, Scientist Who Studied Body Chemistry. *The New York Times*, Jan. 10, 2001. [http://www.nytimes.com/2001/01/10/us/horace-barker-93-scientist-who-studied-body-chemistry.html]
6. Kornberg A. (1991) *For the Love of Enzymes: The Odyssey of a Biochemist.* Harvard University Press, p. 102.

7. Ibid., p. 125.
8. Ibid., p. 135.
9. Ibid.
10. Ibid., p. 142.
11. AK Interview with Sally Smith Hughes, Program in the History of the Biosciences and Biotechnology, Biochemistry at Stanford, Biotechnology at DNAX, Arthur Kornberg, 1997, p. 11.
12. Ibid.
13. Hargittai I. (2002) Arthur Kornberg. In: *Candid Science, II. Conversations with Famous Biomedical Scientists.* Imperial College Press, p. 60.
14. Ibid.
15. Hurwitz J. (2008) Tribute to Arthur Kornberg, Stanford University School of Medicine, January 25, 2008, p. 37 (private publication).
16. Roger Kornberg, personal communication, May 2014.
17. Letter from Arthur Kornberg to Robert Luria, Dec. 24, 1987. Reproduced with permission from the Department of Special Collections, Stanford University Libraries.
18. Roger Kornberg, personal communication, May 2014.
19. AK Interview with Sally Smith Hughes, Program in the History of the Biosciences and Biotechnology, Biochemistry at Stanford, Biotechnology at DNAX, Arthur Kornberg, 1997, p. 72.
20. Ibid.
21. ECF Interview with Randy Scheckman, Oct. 2013.
22. ECF Interview with Robert Lehman, Oct. 2014.
23. Randy Schekman. (2007) Obituary, Arthur Kornberg 1918–2007. *Cell* 131: 637–639.
24. ECF Interview with Roger Kornberg, May 2014.
25. Ibid.
26. Ibid.
27. Ibid.
28. Ibid.
29. Kornberg A. (1991) *For the Love of Enzymes: The Odyssey of a Biochemist.* Harvard University Press, p. 147.
30. Ibid.
31. Ibid., p. 149.
32. Ibid., p. 151.

CHAPTER SIX

The Discovery of DNA Polymerase

"Among my many love affairs with enzymes, the one with DNA polymerase has been by far the longest and strongest," Kornberg wrote in his autobiography.[1]

While grappling with the enzymology of RNA synthesis, Kornberg was also pursuing the biosynthesis of DNA. He searched for evidence of DNA synthesis *in vitro* by measuring the incorporation of radiolabeled thymidine into an acid-insoluble product, a property expected of nucleic acids. He initially obtained radiolabeled thymidine as a gift from Morris Friedkin in the Department of Pharmacology at Washington University, who had a solution enriched with the compound left over from one of his own experiments. Kornberg found the results dubious. After one hour of incubation very little of the radioactive thymidine was converted to an acid-insoluble state. It took almost a year before Kornberg repeated this experiment. This time he used a thymidine solution with three times the amount of radioactivity present in his first trial. Once again only a tiny amount of the nucleoside was incorporated into acid-precipitable material. But he was encouraged to observe that the radioactive product was susceptible to digestion by pancreatic deoxyribonuclease (DNase), an enzyme purified and crystallized by Moses Kunitz at the Rockefeller University in 1950, which specifically degrades DNA. Kornberg was convinced that he had discovered a new enzyme, which he named DNA polymerase, though he readily admitted that without the encouragement of the diagnostic result that DNase yielded he wondered whether he would have had the will to "pursue such a feeble light."[2]

Before even calculating the results of this experiment, Kornberg informed his postdoctoral colleague Bob Lehman about this potentially groundbreaking observation. On hearing this news, Lehman immediately informed Kornberg that he wished to drop his own studies and work with him on DNA synthesis. "I had joined Arthur Kornberg's laboratory in September of 1955 and had begun work on the purification of an enzyme in extracts of bacteriophage T2-infected cells that supported the addition of a hydroxymethyl group to dCMP," Lehman related. "But when Arthur showed me his results I was tremendously excited by the possibility that they represented the first demonstration of DNA synthesis *in vitro* and I asked if I could put my project on hold and join him. He agreed."[3] Lehman switched his research project the very next day and made significant improvements in the enzyme assay. In particular, he observed that thymidine monophosphate was a far better substrate for the putative DNA synthesis than thymidine and that thymidine triphosphate yielded even more persuasive results.[4]

In the spring of 1956 Maurice Bessman, a postdoctoral fellow, and Steven Zimmerman (Kornberg's first graduate student) joined Bob Lehman, Sylvy Kornberg, laboratory technician Ernie Simms and Kornberg in further experiments in their 20-by-20-foot laboratory. "Crowded, excited, sharing good spirits, ideas, and reagents, we produced results the sum of which was far greater than it would have been were we diluted into a larger space separated by walls," Kornberg commented.[5]

"The most complex and revealing insights into the reaction would come from exploring the function of the DNA that I had included in the reaction mixture," Kornberg wrote.[6] In his initial experiments, Kornberg added exogenous DNA to his enzyme reactions in the hope that it would serve as a primer for extended DNA synthesis. He entertained this hope influenced by the work of Carl and Gerti Cori on the growth of carbohydrate chains by glycogen phosphorylase, work that demonstrated that the assembly of a starch-like chain depends on the presence of a preformed chain to elongate. As revealed in a later chapter, DNA replication in living cells does indeed utilize a primer, one made of RNA. But in Kornberg's early experiments the added DNA was acting as a template, a startling realization that Kornberg and Lehman only came to several months later. No other enzyme had ever been described that took its instructions from the substance it was working on. "I never presumed that I would immediately discover a phenomenon so utterly unprecedented in biochemistry; one in which an enzyme is absolutely dependent on its substrate, a template, for instruction," he wrote.[7] In later years the esteemed French molecular biologist and biochemist Jacques Monod, confronted Kornberg with the pointed question: "Arthur, do you appreciate the significance of the enzyme?" "His comment is telling," Kornberg told an interviewer. "He thought that I lacked appreciation of what that enzyme accomplished — an enzyme unique in enzymology.[8]

Kornberg presented his discovery of DNA polymerase at two scientific meetings convened in 1956, separated by a mere few months. The first presentation, in the spring, was at a meeting of the Federation of American Societies for Experimental Biology (colloquially referred to as the Federation Meetings) in Atlantic City, New Jersey, an annual Mecca for hundreds of scientists in the biological sciences. In June of that year, he made a presentation at a symposium on *The Chemical Basis of Heredity* convened at the McCollum-Pratt Institute of Johns Hopkins University in Baltimore, MD. "[A] student today, would be surprised by the uncertainties in my report," Kornberg wrote. "Yet it was clear that we were exuberantly on our way to understanding the replication of DNA."[9]

The published abstract of the work presented in Atlantic City represents the first written documentation on the *in vitro* synthesis of DNA. This was quickly followed by a brief report in mid-1956 that included just two tables, published in *Biochemica et Biophysica Acta*. Entitled "Enzymic Synthesis of Deoxyribonucleic Acid," the two-page article opened with the modest statement: "We have reported (citing the published abstract for the above-mentioned Federation Meetings) the conversion of ^{14}C-thymidine via a sequence of discrete enzymic steps to a product with the properties of DNA."[10]

The discovery quickly attracted widespread attention, including from those then interested in the structure of DNA. In August 1956, Kornberg received a letter from Maurice Wilkins at King's College, London. "It is with the greatest interest that we hear of your experiments on the synthesis of DNA," Wilkins wrote. "We are now finishing the details of our X ray diffraction work on DNA and would be very interested to widen our studies somewhat. I wonder if your work had yet reached a stage where you might be able to produce sufficient material for X ray study?"[11]

In 1957, the proceedings of the symposium convened in Baltimore the previous year saw the light of day. It included a lengthy paper by Kornberg entitled "Pathways of Enzymatic Synthesis of Nucleotides and Polynucleotides," in which he surveyed the work in his laboratory on the biosynthesis of ribonucleotides, deoxyribonucleotides, coenzyme synthesis, the synthesis of RNA — and DNA synthesis by DNA polymerase. In concluding this overview he wrote about DNA polymerase:

> we know relatively little about the enzyme reaction and the nature of the DNA product. The overriding question remains: how is biologically-specific DNA formed? Nevertheless, the relative simplicity of the DNA-forming system described now makes a number of experiments feasible. The enzymatic synthesis of a bacterial transforming factor, once regarded beyond experimental reach, has now become an immediate objective.[12]

Another cogent reason for Kornberg adding exogenous DNA to his early DNA polymerase reactions was the hope that its sheer mass would outweigh any newly synthesized DNA and hence would be preferentially degraded by random nucleases in the reaction mix. Fortuitously, this consideration turned out to provide an additional bonus since the degraded DNA added to the nucleotide pool for DNA synthesis. These nucleotides were activated by enzymes in the extract that used ATP added to the reactions to generate the triphosphate forms of the bases A, G, C and T and the radiolabeled thymidine.

"Going over the laboratory notes of these early explorations in broken cell extracts, I shudder to see the odds we faced and our innocence of them," Kornberg wrote years later.

> The signal we followed was barely detectable, our supplies of the building blocks were scanty and primitive, the techniques for fractionating enzymes were few and blunt, and there were no genetic guidelines for enzymes in the synthetic pathway. Yet my faith in the power

of biochemistry was so great, that having devised an assay, I believed I could force a wedge into this tiny crack and use the hammer of enzyme purification to drive it to a solution.[13]

The phrase "the hammer of enzyme purification" became synonymous with Arthur Kornberg.

Between 1958 and 1961 Kornberg and his laboratory colleagues published eight manuscripts in the *Journal of Biological Chemistry* or the *PNAS*, all graced with the primary title "Enzymatic Synthesis of Deoxyribonucleic Acid," but with varying subtitles. About a month after the first two papers in this historic series were submitted to the *Journal of Biological Chemistry* (*JBC*) in October 1957, Kornberg was shocked and dismayed to be informed that the editor considered both papers to be unacceptable for publication. "There were no objections to the substance of the papers — a description for the first time of the novel triphosphate forms of the DNA building blocks, procedures for their synthesis, and the isolation of an enzyme that polymerized them," he wrote. "Nor were there criticisms of our characterization of the product — except that it should not be called DNA. Some insisted that we call our synthetic product by the accurate but vapid name 'polydeoxyribonucleotide.'"[14] One reviewer whose caustic literary style unambiguously identified him to Kornberg* requested evidence of genetic activity in the synthetic substrate to have it qualify as DNA.[15] Though acutely aware of the importance of such experimental evidence, genetic proof was not something conjured up at the snap of a finger — and in fact was almost a decade away.

After several heated exchanges with the journal editor, Kornberg seriously considered withdrawing the papers from further consideration. In December 1957 he wrote to the incumbent editor of the *JBC*, the celebrated Harvard biochemist John Edsall, communicating his aggravated state of mind.

---

* A piece about Kornberg on the Internet informs: [Kornberg] states that the journal told him that a peer, the noted biochemist Erwin Chargaff, had written an exceedingly sarcastic letter in assessing his findings." [http://bitesizebio.com/articles.arthur-kornberg-biochemist]

> Yesterday I received a letter from the Editors of the Journal, which is unique in my experience. It is derogatory and insulting; in places it may even be malicious. -------------------------- My first impulse on receiving this letter yesterday was to withdraw the manuscripts from consideration by the Journal and to resign from the Editorial Committee (of which Kornberg was a long-standing member). On more sober reflection, I find that I am just as devoted to the journal itself as ever, and I feel an even greater obligation to serve in any capacity that I can to improve its quality and administration.[16]

Edsall read both papers and was unequivocal in his intention to see them in print in the *JBC*. In responding to Kornberg's December 11 letter he wrote:

> I feel very strongly, as the future editor of the *JBC* that I should be extremely disappointed if work of this sort were not to appear in the Journal, but were to be sent somewhere else.[17]

Edsall suggested to Kornberg that he bide his time until the anticipated transition of editorship of the journal was officially consummated. True to his word the two papers appeared in the May 1958 issue of the *JBC*.

Purification of the DNA polymerase identified in extracts of *E. coli* cells was not a trivial task. The enzyme was present in crude extracts in relatively small amounts, even in cells in the logarithmic phase of growth. Fortunately, the biochemistry department at Washington University possessed a fermenter that greatly alleviated this limitation. In later years, a private entity was contracted to produce 100 pound batches of *E. coli* cell paste![18,19]

Kornberg was naturally keen to show that the DNA synthesized by DNA polymerase was a faithful copy of the template. He realized that for this to be true the synthetic product would have to fulfill two criteria. First, there should be a strict equivalence between the amount of the bases adenine and thymine and between guanine and cytosine. In the summer of 1958 this expectation was fully realized. A second criterion expected to be fulfilled was one that would identify the DNA of the particular species used as a template, revealed by the ratio of A-T to G-C base pairs, a ratio that varies from less than 0.5 to near 2.0 for different species. The results were equally satisfying. When replicating DNA from the bacterium *E. coli* in which the ratio of A+T over G+C was known to be 0.97, experiments yielded a value of 1.02. For a bacteriophage called T2, with a known AT:GC ratio of 1.9, experiments yielded a product ratio of 1.92.

"Analyses of our synthetic DNA told us only that within a few percent we could reproduce the overall A, T, G and C composition," Kornberg pointed out. "But it revealed nothing about their sequential arrangement in our product." Together with postdoctoral fellow John Josse and with considerable assistance from Dale Kaiser, Kornberg devised an elegant procedure for determining the frequency with which any one of the four nucleotides in a DNA synthesized by DNA polymerase in the test tube is next to any other. The outcome of these so-called nearest-neighbor analyses was also spectacularly gratifying. "The results were all we could have wished," Kornberg wrote. "For each synthetic DNA there was a distinctive distribution of nearest-neighbor sequences."[20]

Providentially, Kornberg's analyses of nearest neighbor frequencies unambiguously identified the direction of the two chains in a duplex DNA molecule, a fact that could not be addressed by Watson and Crick when they postulated the structure of DNA in 1953. When Kornberg and his colleagues placed nearest neighbor sequences in DNA chains oriented in opposite directions, the abundances of each of the 16 possible nearest-neighbor pairs matched almost perfectly. In contrast when the chains were oriented in the same direction, the abundances of the sequences facing each other revealed no match. "This relationship held for a variety of animal and bacterial DNAs we tested and left no doubt that the enzyme, in its synthetic activity, had taught us a major truth about DNA: the two chains of the DNA double helix run in opposite directions," Kornberg triumphantly stated.[21]

In the spring of 1960 Kornberg shared his nearest-neighbor data with Francis Crick, who responded:

> ------ the base-pairing [data] is very satisfactory. I was very glad to see that in your *Fed. Proc.* note you stressed that this implies that the [DNA] chains run in opposite directions, as this is the only *good* piece of evidence in favour of this feature, apart from the complete X-ray data, and the interpretation of that still leaves much to be desired.[22]

And when Kornberg passed away in 2007, a tribute by Marc Bretscher at the Laboratory of Molecular Biology in Cambridge, UK highlighted his deduction of anti-parallel DNA chains.

> This was an extraordinary piece of chemical evidence, coming from an entirely unexpected quarter: it supported the double helical structure for DNA which had been based largely on X-ray studies and model building.[23]

Nor was this the only beneficial utility of the enzyme DNA polymerase. With the advent of the molecular cloning of genes in later years, DNA polymerase became one of the most useful tools in the cloner's toolkit.

Kornberg was on a roll! For all the disparaging comments and retorts that permeated some quarters, life in St. Louis was more than tolerable in general — and spectacular scientifically. But Kornberg was becoming weary of the question that followed essentially every one of the many invited seminars he presented during the 10 years between 1957 and 1967. "Why have you been unable to replicate the genetic activity of the DNA template?" colleagues insistently asked. As noted in a later chapter, in due course Kornberg did address this question (see Chapter 12).

In concluding this chapter on Kornberg's discovery of an enzyme that replicates DNA it bears emphasis that he was always fundamentally interested in understanding the *enzymology* of DNA replication, not the role of DNA as the harbinger of genes. When in 1997, he was asked by an interviewer: "Am I right that your interest in the synthesis of DNA was not so much because it happened to be the genetic material but because it was a logical extension of your previous work [as an enzymologist]?" Kornberg unhesitatingly replied:

> That's absolutely true. ----- [The enzyme that synthesizes DNA] wherever you find it — and it is found everywhere DNA is made — takes instructions from preexisting DNA and therefore replicates the genetic material. It was not lost on me that this enzyme with these very impressive properties of copying a template was of direct importance in the replication process. But I can't honestly say that I sought that enzyme to solve a biological problem.[24]

Kornberg was always adamant in his view that his interest in DNA polymerase was chiefly motivated by his love of enzymes. "I do want to emphasize that it has been my conviction, and it's the basis of the books I have written on DNA replication, that you have to know the actors to understand the plot," reiterating his thematic view about enzymes and the biochemical reactions they support.

> In this case we were rewarded by seeking out an enzyme that does something as simple as adding one building block to an existing chain and finding that this event required that for building a chain you have all the components that enable the actor, DNA polymerase, to do its job.[25]

Later chapters will reveal that cells contain more than one DNA polymerase and that the enzyme discovered by Kornberg was not the polymerase that replicates DNA *in vivo*. "But," Kornberg pointed out in 1999, "the fact that there are other DNA polymerases doesn't alter the fact that they all carry out the very same basic catalytic activity of matching a template with a nucleotide."[26]

References

1. Kornberg A. (1991) *For the Love of Enzymes: The Odyssey of a Biochemist.* Harvard University Press, p. 154.

2. Ibid., p. 151.
3. ECF Interview with Robert Lehman, Oct. 2014.
4. Lehman IR. (2003) Discovery of DNA Polymerase. *J Biol Chem* 278: 34733–34738.
5. Kornberg A. (1991) *For the Love of Enzymes: The Odyssey of a Biochemist*. Harvard University Press, p. 153.
6. Ibid., p. 155.
7. Ibid., p. 153.
8. AK Interview with Sally Smith Hughes, Program in the History of the Biosciences and Biotechnology, Biochemistry at Stanford, Biotechnology at DNAX, Arthur Kornberg, 1997, p. 21.
9. Kornberg A. (1991) *For the Love of Enzymes: The Odyssey of a Biochemist*. Harvard University Press, p. 155.
10. Kornberg A, Lehman IR, Simms ES. (1956) Enzymatic Synthesis of Deoxyribonucleic Acid. *Biochem Biophys Acta* 21: 197–198.
11. Letter from Maurice Wilkins to Arthur Kornberg, Aug. 20, 1956.
12. Kornberg A. (1957) Pathways of Enzymatic Synthesis of Nucleotides and Polynucleotides. In: WD McElroy and B Glass (eds), *Symposium on the Chemical Basis of Heredity*, pp. 579–605.
13. Kornberg A. (1991) *For the Love of Enzymes: The Odyssey of a Biochemist*. Harvard University Press, p. 155.
14. Ibid., p. 158.
15. Ibid.
16. Letter from Arthur Kornberg to John Edsall, Dec. 11 1957. Reproduced with permission from the Department of Special Collections, Stanford University Libraries.
17. Letter from John Edsall to Arthur Kornberg, Dec. 27, 1957. Reproduced with permission from the Department of Special Collections, Stanford University Libraries.
18. Lehman IR. (2003) Discovery of DNA Polymerase. *J Biol Chem* 278: 34733–34738.
19. Friedberg EC. (2006) The Eureka Enzyme: The Discovery of DNA Polymerase. *Nature Rev Mol Cell Biol* 7: 143–147.
20. Kornberg A. (1991) *For the Love of Enzymes: The Odyssey of a Biochemist*. Harvard University Press, p. 160.
21. Ibid., p. 161.

22. Letter from Francis Crick to Arthur Kornberg. May 10, 1960. Reproduced with permission from the Department of Special Collections, Stanford University Libraries.
23. Bretscher MA. (2007) Professor Arthur Kornberg, Nobel prize-winning Biochemist Whose Research Into Enzymes Helped Unravel the Mystery of DNA. *The Independent,* Saturday, Nov. 3, 2007.
24. AK Interview with Sally Smith Hughes, Program in the History of the Biosciences and Biotechnology, Biochemistry at Stanford, Biotechnology at DNAX, Arthur Kornberg, 1997, p. 62.
25. Ibid.
26. AK Interview with Sally Smith Hughes, Program in the History of the Biosciences and Biotechnology, Biochemistry at Stanford, Biotechnology at DNAX, Arthur Kornberg, 1997, p. 63.

CHAPTER SEVEN

Stanford University Medical School

Prior to the mid-1950s the Stanford Medical School was geographically split. The basic sciences, including biochemistry, which was taught by Herbert Loring and Murray Luck in the Department of Chemistry, took place at the Stanford campus in Palo Alto. However, clinical training was provided at hospitals in San Francisco, a good 30 miles north. The decision by the university to consolidate its medical school entirely at the Stanford campus consumed protracted and often-contentious discussion. In his book *Biotech — The Countercultural Origins Of An Industry,* Eric Vettel summarized the combative atmosphere that preceded the move. "A meeting with the University Board of Trustees in May 1953, just two months after Watson's and Crick's famous discovery, a wake-up call to many academic institutions aspiring to state-of-the-art basic biomedical research, afforded Stanford Provost Fred Terman the opportunity to restructure and reorganize the biological sciences," Vettel wrote[1]

> He was strongly supported by President Wallace Sterling, who spoke forcefully on the topic. "We have a medical school problem," Sterling told the trustees. "Medical education, which is now in a state of flux, is inextricably tied to the basic sciences. ----------------- The key is the relationship of medical education --- to other scientific fields." Then Sterling offered an unexpected and startling solution: "bring the Medical School into the closest possible physical and intellectual relationship to the whole University."[2]

Robert Alway, dean of the Stanford Medical School from 1957 to 1964.

Many in the local communities proximate to the Stanford campus were opposed to the presence of a medical school in their midst that was devoid of a primary mission of delivering health care. "Neither medical training nor patient care was the driving force behind Sterling's proposal to move the university hospital from San Francisco to Palo Alto — and Stanford faculty knew it. So did local communities," Vettel stated.[3] The name of the game was basic research.

Ultimately the Palo Alto City Council followed the lead of former childhood actress Shirley Temple-Black, a long-time citizen of Woodside, a rural town near Palo Alto. Possessed with a long and distinguished history in politics, Temple-Black persuaded the Palo Alto City council to "threaten to veto any of Stanford University's applications for building permits until the university gave some indication that the new campus would provide health care for local communities."[4] Realizing that their attempt to promote the biological sciences in the medical school had seriously split the resident community, the Stanford board of trustees reluctantly endorsed plans for a new 500-bed research hospital on campus. "Thus, in a bizarre turn of priorities, Stanford signed an agreement with the City of Palo Alto to jointly build and operate a community hospital."[5]

Edward Durrell Stone was selected as the architect for Stanford's $21.5 million, 400-bed Palo Alto-Stanford Medical Center. Comprised of three hospital and two medical school buildings interconnected by numerous arches and open walkways, the strikingly beautiful 56-acre site was designed along lines similar to those of Stanford's imposing main quadrangle. A three-story height was maintained throughout and the concrete walls and columns of the center were patterned as a grillwork to simulate the sandstone surfaces of the main university Quad.[6] In later years another building was added to the main complex and a fourth, the Fairchild building, was added as a standalone structure adjacent to the main building. That was in turn followed by the erection of the Arnold and Mabel Beckman Center, a research facility that opened its doors in May 1989 with a $12 million gift from the Arnold and Mabel Beckman Foundation, which covered one-fifth of the nearly $60 million construction and outfitting costs.

The decision to relocate the medical school to the Stanford campus promoted the intention of the Stanford administration to offer the chairmanship of a new biochemistry department to Edward (Ed) Tatum (whose seminar had so inspired Kornberg early in his sojourn at the NIH; see Chapter 2), a faculty member in the Stanford Department of Biology, and move chemistry professors Hubert Loring and Murray Luck into the new department.[7] But when in 1957 Tatum elected to join the Rockefeller University in New York, the chairmanship of the biochemistry department was left in a vacuum.

There were then other forces at the medical school that wanted to see a different outcome for the leadership of the new Department of Biochemistry. In particular, the late Henry Kaplan, head of the Stanford Department of Radiology, and Avram Goldstein, chairman of the Department of Pharmacology, two men with considerable influence concerning the establishment of basic science programs in the new medical school, harbored grander ambitions for biochemistry at Stanford.

Henry Kaplan was then toying with an offer to move to Boston to assume the chairmanship of the radiotherapy department at Harvard University.

When provost Terman caught wind of Kaplan's possible defection from Stanford, he began discussions designed to retain him. But Kaplan protested that he didn't want anything for himself. He wanted to be surrounded by "intellectual playmates!" Identifying the intended biochemistry department as a case in point, Kaplan proffered that he and Goldstein had identified the ideal candidate in Arthur Kornberg.[8]

A world-renowned radiotherapist and radiobiologist who was eagerly sought by a number of medical schools, Kaplan was not especially known for humility — and could be notoriously belligerent when seeking to impose his will. He took it upon himself to aggressively promote Kornberg's recruitment. "A very important person was Henry Kaplan," Kornberg related.

> He was radically different from other radiologists in that he was both biochemically oriented and very skilled clinically. We had known each other at the NIH. He approached me in the spring of 1957, mentioning the rather amorphous situation in biochemistry at Stanford and asked whether I'd be interested. I said I might be; I'd been in St. Louis for four years; a long time in that city and I'd come to California during several summers to get away from the place.[9]

Kornberg didn't want rumors about possibly leaving Washington University to circulate and he never mentioned his early discussions with Kaplan to a soul. "It was with some relief that I didn't hear anything further about Stanford for a good while," he related, "because I was busy with my work and things were going well."[10]

Kaplan had a reputation as a "dean-eater" with limited tolerance for medical school deans who in his opinion failed to live up to his leadership expectations. In her 2010 biography entitled *Henry Kaplan and the Story of Hodgkin's Disease*, author Charlotte deCross Jacobs documented an interchange between Kaplan and Stanford Provost Fred Terman concerning Kaplan's interest in recruiting Kornberg to head a new Department of Biochemistry at Stanford. "'I've watched how Stanford does some of its recruiting," Kaplan stated.

> You wait until you hear that the guy has been invited out to give a seminar at Berkeley so that you don't have to pay his plane fare, and then you invite

him across the bay to the farm, and you talk to him about the sunshine, the climate, and you don't offer him any budget or any space, and then you don't understand why he won't come.

When Terman wearily asked Kaplan what exactly he would wish Terman to do Kaplan responded:

> I'd like you to invite Mrs. Kornberg with Arthur on the first visit. I'd like them to come first class; I'd like to have a car waiting for them at the airport, preferably a convertible. I'd like them to have a suite at Rickey's (a hotel in Palo Alto). ------------------- I want you to promise him every square inch of space and every dollar of budget he asks for, because I know Arthur well enough to know that he won't ask for more than he can use. --------- I'll make it very simple for you. If you recruit Kornberg successfully I'll stay at Stanford. If you fail I'm going to Harvard.[11]

Kornberg was surprised, but not shocked, to receive a formal invitation to consider moving to Stanford University. "There was an invitation to come out to look at the job," he related. "I might even have asked: 'I'm not interested in looking ---- are you people serious?'"[12] The university's response was emphatically affirmative. Not too much later Kornberg and his wife Sylvy visited Stanford at the invitation of Robert (Bob) Alway, dean of the medical school. The visit was highly encouraging. "Not only do we want you, but you're not being looked at; you're looking at us. What can we do to interest you?" he was told. And there would be a new building. "Just plan the space. We'll send the architects out to St. Louis; you can design what you want."[13]

"In the handsome new buildings designed by Edward Durell Stone the biochemistry department was assigned a spacious floor that I could design to my own specifications and the unpopulated department would be staffed with colleagues of my choosing," Kornberg related.[14]

> Now a legitimate biochemist, I could teach and practice my subject without disguise and appoint the physical and organic biochemists whom I very much missed. No longer would I be the muted, junior member of the tradition-bound Executive Committee of Washington University

Medical School. Rather, I would have a large share in policy and faculty selection in the renaissance of science and medicine at Stanford promised by J. E. Wallace Sterling, the President and Frederick E. Terman, the Provost.[15]

In June 1957, Alway communicated with Kornberg in writing. "I am very happy officially to reaffirm our invitation to you to become Professor and Executive of the Department of Biochemistry of the School of Medicine of Leland Stanford Junior University," he wrote. "This letter is further to confirm and extend various points considered in several talks last week. I talked with President Sterling this morning following his return to the west coast last evening, and now find my way entirely clear actively to pursue what I trust is our mutual interest."[16] In an obvious gesture of camaraderie Alway addressed Kornberg as "Dear Art" (an abbreviation of Kornberg's Christian name never otherwise encountered in his voluminous correspondence from others). A hand-written postscript states: "More stuffy associates might shudder at my informal salutation in such a letter — but — simply may I add — the weather continues its usual — delightful!" In a prompt response to Alway's letter Kornberg wrote: "When I left Stanford a little over a week ago, I felt very much inclined to accept this offer and despite some touching and painful experiences during the past week I am still so inclined."[17]

A brisk correspondence followed, including a lengthy letter dated July 1, 1957 in which Alway made mention once again of the pleasing weather in Palo Alto, stating: "the temperature is 74, the sky is blue, and a cool breeze is blowing down the Bay,"[18] climatic and geographical nuances that true to Kaplan's predictions were always among the cards in Stanford's recruiting deck.

Unsurprisingly, word of Stanford's interest in Kornberg soon reached Washington University, which unsurprisingly launched its own full-court press to retain him in his position as chair of the Department of Microbiology. "As soon as it became apparent that I was leaving I was told by the dean that when Cori retired I would become professor of biochemistry," Kornberg stated. "So there were efforts to persuade me and efforts to persuade some of my people to stay — Berg and Cohn — and others."[19]

When in one of his numerous letters to Henry Kaplan during that period Kornberg, perhaps during a fleeting indecisive moment, informed Kaplan of the pressures from Washington University, Kaplan responded with a reassuring and flattering epistle.

> "I am not a bit surprised at the tremendous amount of pressure being put upon you, very understandably, by Cori and all of your other colleagues there, and can only say that I am delighted that you want to come nonetheless," Kaplan wrote. "There can be no question that the intellectual climate at Washington University School of Medicine is far superior to that which we enjoy and only a small number of departments here have really expressed the kind of leadership which is evident in almost every department there. However, as you know, these things have a way of changing almost catalytically and I would look upon your arrival here as a very significant catalyst. Working together, I am sure we can do much to revolutionize the situation without at the same time causing any bitterness or recriminations. I believe that opportunities for making new appointments will be opening up in two or three departments within the next two or three years and having a man with your imagination and judgment of other men on hand to participate in the selection of new department heads would be a major factor in making a happy choice. Within a very few years, by such steps, the School could arrive at a position of considerable strength."[20]

Kaplan, who clearly saw himself as the leading force behind Kornberg's recruitment, constantly updated Kornberg about efforts to bring this intention to fruition.

> "I talked with Bob Alway the other day just before he went out of town again," he wrote. "I don't know whether he is going to be able to write you while on his current trip or whether he will defer this until he gets back. ---------------- I believe that he was able to clear up virtually all the points raised in your most recent letter to him and he discussed some of these items with Avram and me. I really see no obstacle whatsoever and feel that essentially everything you have asked for from Stanford it will be in a position to grant. -------------------- Your coming to Stanford would be a tremendous personal source of gratification for me."[21]

Kornberg ultimately formally accepted the Stanford offer. On July 3, 1957 he informed Alway of his decision.

> Dear Bob,
>
> I received your letter of July 1 and enjoyed its spirit and details. My colleagues and I are delighted with the prospect of going to Stanford and I accept with enthusiasm the offer of Stanford University School of Medicine to become its professor of biochemistry.[22]

On the same day he dropped a brief note to Henry Kaplan.

> Dear Henry,
>
> I have just written to Bob Alway accepting the job. If I've overlooked any important details I have the confidence they will be handled in the same spirit that matters have been considered up to now.[23]

By all accounts Kornberg's negotiations with the Stanford administration were free of any protracted or disconcerting discussions. Indeed, he was genuinely impressed with the university leadership. "We had an impressive administration here at the time," he related in later years. "Wallace Sterling (the Stanford president) had lofty goals and his provost, Fred Terman, was a no-nonsense, uncharismatic, but very effective person."[24]

The only issue that arose concerned Kornberg's relationship to members of the Department of Chemistry who had assumed the responsibility of teaching biochemistry. "The chemistry department, as it was being reorganized, wanted desperately to unload Hubert Loring and Murray Luck, two people in the Department of Chemistry who had been teaching biochemistry," Kornberg related.

> Stanford Provost Terman broached that with me. I said, "No!" They were much older than we were and doing research that was different from what we were doing. I couldn't possibly forgo two appointments; it was going to be a small department of seven or eight people. Terman agreed. I could have found ways in which to involve Luck and Loring more than I did, and I'm sorry I didn't. But I was absolutely right in not including them in the department.[25]

"In the summer of 1957 I had to tell Carl Cori that I was leaving," Kornberg stated.

> I hated anticipating this moment, but it had to be done. When I walked into Cori's open office he said: "Well, what have you decided to do?" And I said, "I've decided I'm going to go to Stanford." It was the only time in our long association that he was irritated, angered. He sputtered and said: "Well, where will you go on vacation?" Gerty immediately calmed him down and said: "Carly, we should have gone to Berkeley when we were offered the opportunity."[26]

Kornberg also regretted terminating his long and productive association with Ernie Simms, who he described as a devoted and effective research assistant. "Disadvantaged by lack of college training due to his black origins, — his roots in St. Louis were too deep for transplanting."[27]

"We were going to be lame ducks in St. Louis for two years," Kornberg ruefully remarked. "When we left St. Louis there was a fair amount of ill feeling. At first there was the pleasure and pride of having a bright young group like ours — and suddenly it was gone. And Kornberg took it with him."[28]

At the very outset Kornberg had determined that if he accepted the chairmanship of the Department of Biochemistry at Stanford, a non-negotiable condition was that he be permitted to invite senior members of his laboratory at Washington University to join him on the faculty. Stanford's response was unequivocal. "Yes, you can bring your whole group to Stanford. We'd love to have them," was Provost Terman's rejoinder. "You won't have to go through months and years of search committees and all kinds of bureaucracy."[29]

On returning to St. Louis from his visit to Stanford, Kornberg communicated the prospects there to members of his laboratory. "The enthusiasm of all my departmental associates to join me ultimately made my decision easy," he wrote.[30] A letter to his future Stanford colleague Avram Goldstein communicated this news:

> I think I may have mentioned to you that my decision to accept the Stanford offer would very likely be concurred in by the rest of the

department, and this has been so. Mel Cohn, Paul Berg, Dave Hogness and Dale Kaiser, who hold staff appointments in our department at present, are eager to continue as a group at Stanford; Jerard Hurwitz and Bob Lehman, who hold special appointments are also happy to move with us.[31]

Bob DeMars was not invited to move to Stanford. "He was still in the army, and I didn't invite him — because he was a little bit of a maverick — and I had Dale Kaiser," Kornberg commented. "I thought in a small department there wasn't room for an additional appointment in the area of virology."[32]

Jerry Hurwitz also did not relocate to Stanford with the rest of the Washington University faculty. "That was very traumatic," Kornberg told interviewer Sally Smith Hughes.

> Jerry and Paul [Berg] grew up together as students. Jerry is more combative and blunt and was doing very similar things to what Paul and I were doing. It was really sort of good-natured competition. [But] I had to choose between him and Paul, and I chose Paul. That precipitated a "nervous breakdown", as we called it then. Jerry was utterly shattered by that and I've had that on my conscience. But we're very good friends now and have been for many, many years.[33]

Years later, whilst in the midst of deciphering the intricacies of DNA replication in his Stanford laboratory, a member of Kornberg's research group, William (Bill) Wickner, was the brother-in-law of Sue Wickner, a graduate student in Hurwitz's New York laboratory. Kornberg commented that frequent telephone exchanges between Bill and Sue encouraged the Hurwitz laboratory "to adopt the same course we were taking and to make it a race,"[34] a mentality for which Kornberg expressed unqualified disdain. Jerry Hurwitz's "broad interests and ambitions, combined with his experimental skill, knowledge, and accomplishments, have left him few territorial inhibitions," Kornberg wrote.[35]

Hurwitz offers a different recollection of what transpired around that time. In a nutshell, he viewed Kornberg's protective attitude concerning research on DNA replication to be unreasonable. "The only issue that

I had in the lab at Washington University was that everything that I wanted to do Arthur did as well," Hurwitz stated.

> And I thought: "there is no future in this". Once I was interviewing a prospective graduate student and told him that I was especially interested in how DNA is degraded. When I told him this his response was that he believed that Arthur wanted to work on that as well. Eventually I realized that if I was going to survive as an independent investigator I had to go elsewhere.[36]

In 1958 Hurwitz moved to New York where he joined the microbiology department at New York University School of Medicine under Bernie Horecker's chairmanship. "At that time, influenced by work I carried out in St. Louis, I decided to focus on RNA synthesis," he stated.[37] Hurwitz's efforts were rewarded by contributions that he made to the study of RNA polymerase, an enzyme that he independently discovered in bacteria shortly after Samuel Weiss and Leonard Gladstone reported the discovery of RNA polymerase in extracts of mammalian cells.[38] As for his differences with Kornberg — they are indeed long forgotten.

In the context of his clash with Hurwitz, Kornberg expressed a general distaste for competition in science. "It is no fun for me to strive to do something that is likely to be done just as decisively by someone else at about the same time," he wrote, venturing the opinion that he "would rather work on one of the many problems around the core of my interests and competence, one that is not likely to be solved soon by others."[39]

****

The physical move to Stanford was meticulously planned. "In 1957, the Stanford architects came to St. Louis and we laid out the plans, and everybody was now involved in planning to move," Kornberg related.

> In the period in which we were examining the blueprints, we determined what we'd need in terms of apparatus and began making applications for research grants. Especially in the last year, 1958–59, we spent a lot

of time on the logistics of the move; how we would move perishable materials, where they would go, the furniture — and whatnot. [40]

Subsequent to his decision to invite members of his St. Louis laboratory to join his Stanford faculty Kornberg recruited Robert (Buzz) Baldwin from the University of Wisconsin. When asked how he acquired the nickname "Buzz" Baldwin related that his sister, three years his senior, had trouble saying "brother" when she was small, referring to him as "bruzzer" instead, and subsequently as bruzz — and then buzz! "In my first-grade class at school there were four Roberts and since I did not want to be called Bob, I stuck with Buzz!"[41]

Baldwin and Kornberg first met while attending a biophysics conference in Boulder, Colorado prior to Kornberg's arrival at Stanford in 1959. "In my opinion, molecular biology was about to take off and I was sure that Arthur's department was a place where this would happen," Baldwin related.

> He told me that his microbiology department at Washington University was poised to move to Stanford and he wanted to know whether I was interested in being interviewed for a position. I gave a seminar in St. Louis and soon after that departed to pursue a sabbatical in Copenhagen. In due course I had a letter from Arthur asking if I was interested in a faculty position at Stanford. I quickly wrote back and responded affirmatively. I don't think I even bothered to ask what the salary would be. I was tenured at the University of Wisconsin and I don't think I even asked about tenure at Stanford. As it turned out I was initially not tenured there, even though I came as an Associate Professor.[42]

"Baldwin was identified as a young man of great promise and considerable achievement," Kornberg related. "He had been a Rhodes Scholar and received a degree from Oxford. To my surprise he was eager to come to Stanford. He was appointed in the new Department of Biochemistry to provide us with physical biochemistry, which clearly was needed for a proper department."[43] In late May 1959 Kornberg informed Baldwin: "We're packing up now. The first wave will depart from St. Louis next week — Berg, Cohn, Kaiser and Hogness (if he recovers

Robert (Buzz) Baldwin.

from a pulmonary infection). My family will leave on June 9."[44] Baldwin's arrival at Stanford was coincident with that of Kornberg and his crew.

"Arthur was very much a father figure for me; comparable to my true father, who was also a scientist," Baldwin related.

> After meeting and talking with Arthur I decided that the most meaningful work I could do was to undertake a physical study of DNA replication in collaboration with him. But he never put his name on any of the papers that emerged. He really wanted it to be clear that it was my work. I got to know Arthur extremely well based on that collaboration. He could be a little overbearing at times. But that was about the only character flaw I ever picked up on.[45]

Kornberg lost no time in informing the prominent geneticist Joshua Lederberg, then also at the University of Wisconsin, about his decision to move to Stanford.

> As you probably know, my visit to Stanford materialized into a decision to accept their offer. Aside from the obvious climatic and geographical attractions, I think there will be better opportunities for graduate student teaching and a degree of intimacy between the medical school and university that will minimize trade school behavior. -------------------- It's

awful leaving Washington University; the traditions of scholarship are well developed here and we have a number of good friends.[46]

News of Kornberg's impending move to Stanford ultimately persuaded Lederberg, who according to Kornberg was "sought be everybody,"[47] to accept an offer to join the Stanford faculty to head a new Department of Genetics in 1958, a year in which he brought Stanford Medical School its first Nobel Prize for discoveries concerning genetic recombination and the organization of the genetic material of bacteria. "His preference then, and in subsequent years, was to be a member of the biochemistry department, but I felt his leadership in genetics deserved a larger stage," Kornberg wrote.[48] In his 1997 interview with Sally Smith Hughes, Kornberg expanded on his views about Lederberg. "The Department of Genetics was created to give Joshua Lederberg a basis for operations," he stated.

> He is a genius, but he couldn't pay attention for any length of time to deal with things like personnel and management of resources. He saw very early that genetics was chemistry, but he wasn't prepared to change what he was doing. He thought I would do it. I invited him to do things with me, but he didn't have the stomach for it. He wanted to be in the biochemistry department at the time that he came here. I saw that — how should I put it politely? — it wouldn't be a good fit. His activities were so varied, and he was such an important personality that I didn't think he would fit into this modest family atmosphere. I was right.[49]

There being no suitable space to immediately house a new genetics department, Lederberg and a fledgling group of faculty initially camped out in space lent by the Department of Biochemistry. But the two departments operated differently. "People in the Department of Biochemistry were extremely interactive," Paul Berg related. "Josh's philosophy on the other hand was essentially 'you're on your own'. The anticipated coalescence between the Department of Genetics and biochemistry never really materialized."[50]

Lederberg did not disagree with this assessment. In an introduction to the interview that Kornberg granted Sally Smith Hughes in 1997, Lederberg stated: "I would have indeed preferred joining his new biochemistry department; but we have differences in how (or whether) we voice a philosophy of science that he may have been wise to foresee from the point of our first meeting."[51]

Kornberg's impending move to Stanford piqued the interest of other prominent biologists around the country. Notably, Charles Yanofsky, then at Western Reserve University in Cleveland, Ohio, elected to join the Stanford Department of Biological Sciences. There, among other achievements, Yanofsky made the important observation that the nucleotide sequence in DNA is colinear with the amino acid sequence of the protein it encodes. Over the years, Yanofsky cultivated enduring personal and professional relationships with both Kornberg and Paul Berg. In a letter to Henry Kaplan, Kornberg also spoke highly about Seymour Benzer. Benzer effectively utilized mutants of bacteriophage T4 to generate a fine structure map of one its genes at the single nucleotide level, thereby providing the first evidence that the gene is not an indivisible entity, as previously believed, and that genes are linear. "I have it on pretty good authority that Benzer has turned down the Hopkins offer and is returning to Purdue this coming fall," Kornberg wrote in early 1958. "I assume he has also turned

Joshua Lederberg, chair of the newly created Department of Genetics at Stanford University in 1958.

down the Harvard offer ——————————. I think this makes it more hopeful that he will be receptive at a later date to an offer from Stanford."[52]

During the period prior to Kornberg's move to Stanford, the university was seeking a new chairperson in the Department of Chemistry. In the midst of his extensive correspondence with Avram Goldstein and Henry Kaplan during his waning days at Washington University, Kornberg alluded to the possibility that Har Gobind Khorana, then at the organic chemistry section of the British Columbia Research Council, Vancouver, Canada, whose laboratory would later stun the world with the synthesis of a complete tRNA gene in the test tube, had an interest in matriculating to Stanford to lead the Department of Chemistry. In mid-1957 Khorana wrote to Kornberg stating:

> I would be greatly interested in considering the possible move to Stanford for a variety of reasons, one important one being of course the possibility of being close to you and your group. In this connection a joint appointment in the departments of chemistry and biochemistry appeals to me greatly.[53]

Kornberg lost no time in communicating this encouraging news to Avram Goldstein.

> With respect to the bioorganic area, Gobind Khorana is very enthusiastic about a joint appointment in chemistry and biochemistry.[54]

But in the end Khorana, who shared the Nobel Prize in Physiology or Medicine with Marshall Nirenberg and Robert Holley in 1968, did not come to Stanford.

While still in St. Louis, Kornberg was unhesitant in volunteering his opinion about some of the names that he heard mention of with respect to the chairmanship of the Department of Chemistry at Stanford. A letter to Henry Kaplan voiced his disappointment.

> Mel Cohn mentioned to me yesterday that you had dropped Libby's* name as a possible candidate. It frightened me. I was expecting

---

*Willard F. Libby was a physical chemist and specialist in radiochemistry, particularly hot atom chemistry, tracer techniques, and isotope tracer work. [http://www.nobelprize.org/nobel_prizes/chemistry/laureates/1960/libby-bio.html]

that bright young, active research workers rather than political figures were on the list of candidates.⁵⁵

When he ultimately settled in at Stanford, Kornberg was in fact actively drawn into the recruitment of new faculty in the Department of Chemistry, "which, with the selection of William Johnson as chairman and then Carl Djerassi, Paul Flory, Henry Taube, Harden McConnell, and Eugene van Tamelen, vaulted into national prominence."⁵⁶

"In June 1959, twenty-two people — and five children set out for the promised land," Kornberg wrote in his autobiography.⁵⁷

> Having escaped to California from St. Louis for several summers, Sylvy and I were keen to live permanently in that most agreeable climate and geographical setting, one which would also be attractive to our children, future students, and colleagues. Instead of remaining in the "Gateway to the West", we might now be housed in the manor itself.

Predictably, the move was not without its share of hiccups. To Kornberg's dismay, his introduction to his new department mirrored some of the logistical frustrations he endured when he moved to Washington University. Notably, the physical construction of the new department was not complete. Well known for his intolerance of roadblocks and hindrances, Kornberg was unhesitant in letting his displeasure be known. In short order, a crew of medical school workmen were transferred to the biochemistry department on the third floor of the medical school.

****

The eager assistance that Henry Kaplan lent to recruiting Kornberg to Stanford forged a warm relationship between the two departmental chairs, both on and off the Stanford campus. Kornberg twice nominated Kaplan for the prestigious Borden Award in the Medical Sciences, an accolade established by the Borden Foundation and administered by the Association of American Medical Colleges in recognition of outstanding clinical or laboratory research. "Arthur Kornberg enjoyed dinner parties at the Kaplans', where he and his wife met artists, writers, and political scientists," Jacobs wrote in her biography of Henry Kaplan.

> The Kaplans looked forward to picnics with the Kornbergs and they regularly shared Thanksgiving dinner. Henry spent weekends house hunting until he found the perfect home for the Kornbergs. Arthur and Henry were more than just colleagues; they became devoted friends. They argued about art and baseball, and at scientific meetings they stayed up half the night talking. Life at Stanford had turned out just as Henry had hoped — he was surrounded by 'intellectual playmates' at work, blessed with a group of invigorating, loyal friends.[58]

"But in time, the wonderful esprit de corps eroded," Jacobs wrote.

> Faculty performance did not match expectations; personalities that had formerly been stimulating became grating. ---------------------- And the fastidiousness and authoritarian stance of his friend Arthur Kornberg was beginning to irritate Kaplan too. He felt that winning the Nobel Prize gave credence to Kornberg's opinions on any subject. Kornberg was quoted in newspapers and interviewed on television about topics as diverse as the country's economic destiny, anti-intellectualism and the buffoonery of creationists.[59] Leah Kaplan (Henry Kaplan's wife) recalled one evening when Henry and Arthur were arguing about politics. "Arthur", Henry said in frustration, "when it comes to biochemistry I'll get on my hands and knees before you, but when it comes to politics you're as ignorant as anyone else."[60]

Jacobs noted that at one time Kaplan "ridiculed Kornberg publicly, contending that it was only through clinical income from the Department of Radiology that Stanford was able to indulge in a brand new biochemistry department."[61] Roger Kornberg remembers that time well.

> Arthur was a partner with Henry in those days during the creation of what was certainly the most inspired and strongest institute of scientific medicine in the country, something that he and Avram Goldstein undertook together. It really began with Henry and Avram Goldstein

and fostered a strong motivation to recruit my father. Arthur took over the leadership of that mission, which went far beyond the medical school, to include physics and chemistry. And at that time both Henry and Avram were very appreciative of Bob Alway, the dean of the medical school, who supported this notion.[62]

But true to his reputation as a "dean-eater," Kaplan's support of Alway waned and he began pressing for Alway's resignation. When Alway resigned his position in 1964 "many thought that Kaplan [and Goldstein] had pushed him to the point of exhaustion."[63] "When Kaplan got wound up he became merciless," Jacobs wrote.

> At Executive Committee (a committee of the medical school chairs) meetings he stood up, paced around, yelled, and waved his fist. Increasingly caustic, he didn't seem to care what anyone thought. --------- "There were shouting matches," Kornberg said, "that were inappropriate and unbecoming for people of intelligence and stature."[64]

Kaplan and Kornberg were both members of a search committee for Alway's successor. The committee decided in favor of Robert (Bob) Joy

Henry Kaplan, Chair of the Department of Radiology at Stanford University.

Glaser, an experienced medical administrator who had played key roles at the University of Colorado School of Medicine and the Harvard Medical School. According to Jacobs,

> Kaplan wasn't opposed to Glaser's appointment. They shared similar views on the requisites for a first-rate academic medical center. Each thought the other was acquiring an ally in his cause. [But] what Kaplan didn't appreciate was the extent to which Glaser planned to centralize control. -------- And what Glaser didn't realize was how difficult it was for Kaplan to compromise when he felt strongly about something.[65]

Robert Glaser had come to Stanford full of enthusiasm, with plans to build an outstanding medical center. He was prepared to bring all his creative talents and energies to bear. A bitter confrontation erupted even before Glaser's arrival, one recounted in Charlotte Jacobs' biography in lurid detail. "The issue at stake was the Medical Service Plan, which determined disbursement of professional fees," she wrote. When the medical school first moved to Palo Alto there was no plan to manage the financial aspects of the medical practice.[66] But when arguments erupted about the distribution of profits Bob Alway, Glaser's predecessor, asked Herbert Abrams, a radiologist known for his proficiency with financial aspects, to devise one. "The Medical Service Plan had been in effect for almost two years when Glaser was offered the deanship," Jacobs related. "During his interviews, Abrams [had] explained Stanford's practice plan and its faculty endorsement and Glaser [had] assured him that he would not change this arrangement until he had worked with it for at least a year. But after further reflection Glaser came to the crucial understanding that Stanford's practice plan would impose significant restraints on his ability to implement new programs in his capacity as dean.[67]

Glaser wrote to the University President Wallace Sterling, "proposing a plan whereby all professional fees would be transferred to the dean, who would use this income to benefit the whole school, not selected high-earning departments."[68] The medical school Executive Committee accepted his proposal at a meeting that Henry Kaplan missed, being then away from Stanford. When Abrams learned of these negotiations he was livid. Believing that Glaser had gone back on his word not to change the faculty practice plan, he wrote an inflammatory letter to Glaser expressing serious reservations about his plan and implied that the faculty had legal grounds on which to contest it. When word of this letter reached Stanford President Sterling he contacted Abrams wanting to know why he had written such a letter and immediately contacted members of the search committee (of which Kornberg was a member) — who were outraged. "They didn't believe Abrams had acted alone and suspected Kaplan's hand in the affair," Jacobs wrote.[69]

The next day Abrams met with the Dean's Selection Committee.

> The Saturday morning quiet of the Medical School was accentuated by the silence he encountered when he entered the conference room. Kornberg [and others] were seated around a large table. Arthur Kornberg looked stern, with a crease between his eyebrows, his lips pressed tightly together. He opened the interrogation by asking Abrams to explain why he had written to Glaser. Kornberg accused Abrams of purposely interfering with their work. Incensed, Abrams replied, "I have never come across a more self-righteous prick than you." No one spoke.[70]

Abrams abruptly left the room. When Kaplan returned from his travels "he wasn't brooding over the financial plan or the hiring of Robert Glaser. The crux for Kaplan was broken loyalty. Henry Kaplan [felt] betrayed by his colleagues and he would never forgive them."[71]

On July 1, 1965 Glaser officially became Dean and Vice-President for Medical Affairs at Stanford. "He had visions of building a school to surpass all others," Jacobs wrote. "At the first meeting with the Executive Committee he outlined his long-range plans and the chairmen assured him of their support." Glaser was confident that the controversy over the Abrams letter would soon become a distant memory. Before long, however, he ran up against Henry Kaplan![72] For a while, the two related in a cordial manner. But determined that the new dean would not control him or the future of his department, Kaplan, who adamantly resented Glaser's authority over his budget, began asking for more and more resources. Glaser in turn considered Kaplan's stated needs to be unreasonable.[73] Robert Glaser had come to Stanford full of enthusiasm, with plans to build an outstanding medical center. He was prepared to bring all his creative talents and energies to bear. Five years later, he gave up."[74]

"When Henry Kaplan became dissatisfied with Glaser my father was horrified and he came to understand that Kaplan would not have patience with anyone in the role of dean," Roger Kornberg related. "My father was highly supportive of Glaser and in the end he and others faced Henry down. Henry never forgave him for that. That was the rock on which the ship foundered."[75] "Years later, when Arthur Kornberg tried to reestablish his friendship [with Kaplan] he was rebuffed," Charlotte Jacobs wrote.[76] "Kaplan became embittered, and he inadvertently ended up stymieing the growth of the school he loved so dearly."[77]

In 1983, Kaplan developed a persistent cough that he insistently self-diagnosed as allergic in origin. Prodded by his wife Leah to have his cough professionally attended to he eventually did so. As poignantly documented in her biography, Charlotte Jacobs wrote:

> The next evening, Henry came home looking exhausted. Leah was awaiting him at the front door. "Well," she said as soon as he entered the house, "Did you get an X-ray today?" He stood in the foyer, looking down at the floor. "Yes," he replied. After a prolonged silence, he spoke in a barely audible voice. "I have lung cancer."[78]

Henry Kaplan died in 1984 at the age of 66.

****

1959 was a banner year for Arthur Kornberg. The month of May yielded an extended tour of the Soviet Union, followed almost immediately by the move to Stanford. In addition to a welcome relief from the infernal summer heat in St Louis, October brought with it the announcement that many scientists secretly (or not so secretly) yearn for. In early October of every year (close to the date of Alfred Nobel's birth date on October 10), the Nobel Prize Foundation announces the names of the Nobel Laureates for that year. In 1959, this announcement included Kornberg and his former mentor Severo

Ochoa as winners of the Prize in Medicine or Physiology. "The week of events in Stockholm in December, coupled with stops in Europe before and after was for my family and me the best party of our lives," Kornberg wrote in his autobiography.[79] The congratulations that poured in from around the world included a note from Francis Crick who wrote: "As you may (or may not!) know, you are my favorite biochemist ---------."[80]

Kornberg's accolade made him the second Nobel Laureate at Stanford Medical School. A third prize to the medical school went to Paul Berg in 1980, a fourth and fifth to Kornberg's son Roger and to Andrew Fire in 2006, and a sixth to Tom Sudhof in 2013, who though at Stanford at the time of the award had executed his formative work at the University of Texas Southwestern Medical Center at Dallas. No other medical school in the United States, perhaps the world, can claim such distinction.

The phrase "Nobel Prize" is a household one — even among those with no relationship to the world of academia. Deservedly or not the accolade is without question the most exalted in the world of academic recognition. Winning a Nobel has elevated some to the very pinnacle of gratification and pride — and being overlooked has correspondingly deflated the ego of more than a few. Kornberg touched on this topic in one of his interviews and revealed how this highly secretive roll of the dice plays out each year. "I know a little bit about the workings because I've had a very close friend who has been involved; Peter Reichard," Kornberg related.[81]

> Nominations are solicited from a huge number of different people and then are winnowed to a small number. During the summer preceding the awarding of a prize committee members take home thick folders and go over them in great detail. A nomination may incubate for many years. ----------- The discovery has to be at a moment when the spotlight is in focus on the field in which the discovery is made. If time passes and the spotlight shifts, a great discovery can be ignored. Then too, the committees are very small and a vocal influential member can have a disproportionate influence pro or con. He can say, "Look, forget this person," or, "let's give this person the most serious consideration."

Or, "If we give it A, B and C, how about D? D is deserving. Well, we can't give the prize to four individuals. So let's give it to A or not give it at all." There's a lot of politics of that kind, but not generally driven by ethnic, geographical or national interests.[82]

"But largely, the prize hasn't suffered from politics of prejudice. Anytime you have people, you have politics," Kornberg continued.

The Nobel Prize committee in chemistry has been vilified and excoriated by my colleagues at Stanford in chemistry who will say: "We've never heard of this guy. He's a biochemist." There are vicious letters from chemists to the committee. Well, the committee is made up of colleagues, and they are not immune to criticism. So the next time they're not going to do something that controversial. They'll play it safe. They'll give it to an organic chemist who is well known and deserving. Within a group of five people there is bound to be somebody who is more vocal, influential than someone else, and very persuasive pro or con. You have to remember that the decision didn't come down from the heavens; five guys were working on it, with all kinds of pressures and persuasions and whatnot, and have to select somebody.[83]

The secrecy surrounding the announcement of the Nobel Prizes notwithstanding, some of the excitement and surprise for Kornberg had evaporated the previous day. While at the NIH as a visiting seminar speaker in October 1959, Kornberg resided with his long-time friends Celia and Herb Tabor at their home on the NIH campus. During the visit Tabor informed Kornberg that he had received phone calls from reporters inquiring about him. A day or two later when Kornberg and his son Roger (who at the age of 12 had accompanied his father on that trip) deplaned at San Francisco airport, they were engaged by reporters. Kornberg was aware that the names of Nobel contenders were sometimes "dropped" by the media before the official announcement, but he protested that he gave these inquiries no serious attention. At about 5:00 am the following morning he and his family were awakened by a phone call from a newspaper in the East informing him of the glad tidings. "I felt elated, of course, and surprised, but not shocked," he wrote in later years.[84] His wife Sylvy, naturally as excited as her husband at the news, is said to have playfully told a newspaper reporter: "I was robbed!"[85]

Kornberg's elation was undeniably multiplied by sharing the prize with his esteemed former mentor Severo Ochoa. As mentioned in Chapter 5, in 1955, Marianne Grunberg-Managⓞ and Ochoa had reported the isolation of an enzyme called polynucleotide phosphorylase that catalyzed the synthesis of (poly)A from ADP. The awards to Kornberg and Ochoa were formally announced by the Nobel Foundation *"for their discovery of the mechanisms in the biological synthesis of ribonucleic acid and deoxyribonucleic acid."*[86]

In a brief speech offered during the course of one of the many functions that highlight Nobel week, Kornberg was succinct — but gracious and eloquent. "In accepting this award, it seems obvious that I share it with my wife whose guidance and contributions were most often found between the lines of my publications," he stated.

> I share it with Severo Ochoa, who taught me enzymology. I share it with my colleagues in New York, Bethesda, Saint Louis and Stanford, and with the whole international community of chemists, geneticists, and physiologists who are truly responsible for the progress in nucleic acid biochemistry cited today. With all humility, I salute this community of scholars whose aspirations go far beyond personal motivation and national boundaries.

Kornberg's sense of humor shone through as well. "It is a matter of great pride for the schools and cities to which the Award recipient is tied," he stated. "For example, I received a very nice letter from one of my college professors in which he offered to raise my mark in biochemistry!"

Though he was fully deserving of a Nobel Prize for his scientific contributions in general, the award to Ochoa was somewhat tainted by the fact that polynucleotide phosphorylase, the enzyme discovered by him and his collaborator Marianne Grunberg-Manago, was in fact mechanistically not a polymerase. The enzyme could link a few nucleotides together, but the reaction was highly reversible and it later became clear that polynucleotide phosphorylase typically catalyzes the breakdown of RNA, not its synthesis.[87] Nonetheless the enzyme was extraordinarily useful. Almost immediately, Marshall Nirenberg and his colleague J. Heinrich Matthaei put it to use to generate the first three-nucleotide RNA codons, which coded for the amino acid phenylalanine. This first step in cracking the genetic code entirely depended on the availability of Grunberg-Manago's enzyme.[88] It wasn't until 1960 that Jerry Hurwitz and independently others discovered the RNA polymerase that uses DNA as a template for the synthesis of what came to be known as messenger RNA.[89] But Kornberg staunchly defended the appropriateness of Ochoa's prize, opining that he deserved it for many other discoveries he made that were germinal and significant. The issue of appropriateness might also be raised with respect to Kornberg's award which, though also richly deserved, was for his discovery of a DNA polymerase that turned out not to be the enzyme that replicates DNA in living cells. But in truth one surely cannot fault the Nobel committees for not having infallible crystal balls!

Kornberg protested that he was not always enamored with the reality of being a Nobel Laureate. "You're instantly a sage, someone to be quoted as taking a position on issues in which you're not qualified — social, political and economic issues," he stated.

> And that's uncomfortable. ------------------ And then you associate with other Nobel laureates, a little elite group. And again, I'm not so comfortable with that. It's a coterie. These people have been anointed, and I'm anointed too, and so it's a little club. There was a Nobel Prize jubilee four or five years ago (circa 1992). I didn't enjoy it; a lot of people strutting around and wanting to be regarded as important.[90]

"The Nobel Prize has not altered my life, nor did it totally disrupt my routine even the very first day," Kornberg wrote in his autobiography.

> Of course Sylvy and I were alerted for the big champagne celebration given that afternoon by Leah and Henry Kaplan, which many university notables and friends attended. But we were among the last to arrive because we had to collect one of the children from his music lesson after school![91]

Arthur Kornberg receiving the Nobel prize in Physiology or Medicine in 1959.

Kornberg deployed his stature as a Nobel Laureate for causes he deemed appropriate; political and otherwise. In 1975, he lent his voice to a group of laureates in support of Andrei Sakharov, the Soviet advocate of democratization and human rights who had been denied permission to travel to Sweden to accept the Nobel Prize in physics. Kornberg was among 33 Nobel Prize laureates to send a cable to Soviet President Nikolai V. Podgorny, asking him to permit Sakharov to receive the prize.[92] In late 1972, Kornberg took the California State Department of Education to task over the issue of teaching creationism in schools. A letter co-signed by his faculty colleague Dave Hogness stated:

> We respectfully request the opportunity to appear before you on November 9, 1972 in order to plead that Creation Theory be excluded from science textbooks used in this State. ---------------- We appeal to you out of a profound concern. We believe that an extreme disservice will be

rendered to the education of our youth and to society if California were to use textbooks in biology that include concepts of the origin of man which are at odds with evolutionary principles.[93]

Kornberg emphatically reemphasized these sentiments in an address to the California Medical Association in March 1981. "More than a century after Darwin and Huxley and a half-century after the Scopes trial, creationists are alive and kicking," he told his audience.

> We had Scopes II in Sacramento a week ago. Rev. Jerry Falwell (the renowned American evangelical Southern Baptist pastor, televangelist and conservative political commentator) and his Moral Majority are now working on a national scale for legislation requiring that creation as portrayed in the bible be taught in the public schools. In the Sacramento trial, the Attorney General of California was said to have won his case in defending the State Board of Education against a suit to offer creation as an alternative to evolution in science classes. Yet in his ruling, the judge ordered the State Department of Education to instruct the local school boards and textbook publishers to avoid the error of making evolution theory an official dogma, taught as if it is beyond dispute. What monstrous nonsense.[94]

Not too much later Kornberg received an invitation to participate in a debate on evolution sponsored by Jerry Falwell. His response was: "There simply is no more reason to debate evolution than gravitation or atomic theory."[95]

In the final analysis concerning the coveted Nobel award, Kornberg had this to say to interviewer Sally Smith Hughes in 1997:

> It is not widely appreciated that the choice is not made on the basis of productivity over a number of years, but rather for a discovery. How else to justify that someone who has done nothing of significance before or after the cited discovery is given such recognition? The Nobel Prize has a lottery quality to it. Someone can go into the lab and say: "If this experiment works I could get a Nobel Prize for it." It may be one in a million, like a lottery. There is glamour attached to something that can be reached by almost anyone.[96]

In addition to the Nobel Prize, Kornberg earned a clutch of other notable honors, including the National Medal of Science in 1979, the Cosmos Club Award in 1995 and the Gairdner Foundation Award in the same year. He served as president of the American Society of Biological Chemistry in 1965 and held memberships in the National Academy of Sciences, the Royal Society of London and the American Philosophical Society — among other august professional organizations. He was awarded honorary degrees from 12 universities. Perhaps the most glowing tribute to Kornberg was from his alma mater, the University of Rochester. In September 1999, grand opening festivities marked the official opening of the new Arthur Kornberg Medical Research Building and Aab Institute of Biomedical Sciences at the University of Rochester Medical Center.

References

1. Vettel EJ. (2006) *Biotech: The Countercultural Origins of an Industry.* University of Pennsylvania Press, Philadelphia, p. 54.
2. Ibid.
3. Ibid.
4. Ibid.
5. Ibid., p. 55.
6. Friedberg EC. (2014) A Biography of Paul Berg, *The Recombinant DNA Controversy Revisited.* World Scientific Publishing Co., Singapore, p. 93.
7. AK Interview with Sally Smith Hughes, Program in the History of the Biosciences and Biotechnology, Biochemistry at Stanford, Biotechnology at DNAX, Arthur Kornberg, 1997, p. 33.
8. DeCross Jacobs C. (2010) *Henry Kaplan and the Story of Hodgkin's Disease.* Stanford University Press, Stanford, California, pp. 127–128.
9. AK Interview with Sally Smith Hughes, Program in the History of the Biosciences and Biotechnology, Biochemistry at Stanford, Biotechnology at DNAX, Arthur Kornberg, 1997, p. 22.
10. Ibid.
11. DeCross Jacobs C. (2010) *Henry Kaplan and the Story of Hodgkin's Disease.* Stanford University Press, Stanford, California, pp. 127–128.

12. AK Interview with Sally Smith Hughes, Program in the History of the Biosciences and Biotechnology, Biochemistry at Stanford, Biotechnology at DNAX, Arthur Kornberg, 1997, pp. 22–23.
13. Ibid., p. 23.
14. Kornberg A. (1991) *For the Love of Enzymes: The Odyssey of a Biochemist*. Harvard University Press, p. 181.
15. Ibid.
16. Letter from Robert Alway to Arthur Kornberg, June 13, 1957. Reproduced with permission from the Department of Special Collections, Stanford University Libraries.
17. Letter from Arthur Kornberg to Robert Alway. June 18, 1957. Reproduced with permission from the Department of Special Collections, Stanford University Libraries.
18. Letter from Robert Alway to Arthur Kornberg, July 1, 1957. Reproduced with permission from the Department of Special Collections, Stanford University Libraries.
19. AK Interview with Sally Smith Hughes, Program in the History of the Biosciences and Biotechnology, Biochemistry at Stanford, Biotechnology at DNAX, Arthur Kornberg, 1997, p. 31.
20. Letter from Henry Kaplan to Arthur Kornberg, June 24, 1957. Reproduced with permission from the Department of Special Collections, Stanford University Libraries.
21. Ibid.
22. Letter from Arthur Kornberg to Robert Alway, July 3, 1957. Reproduced with permission from the Department of Special Collections, Stanford University Libraries.
23. Letter from Arthur Kornberg to Henry Kaplan, July 3, 1957. Reproduced with permission from the Department of Special Collections, Stanford University Libraries.
24. AK Interview with Sally Smith Hughes, Program in the History of the Biosciences and Biotechnology, Biochemistry at Stanford, Biotechnology at DNAX, Arthur Kornberg, 1997, p. 22.
25. AK Interview with Sally Smith Hughes, Program in the History of the Biosciences and Biotechnology, Biochemistry at Stanford, Biotechnology at DNAX, Arthur Kornberg, 1997, p. 33.

26. AK Interview with Sally Smith Hughes, Program in the History of the Biosciences and Biotechnology, Biochemistry at Stanford, Biotechnology at DNAX, Arthur Kornberg, 1997, p. 31.
27. Kornberg A. (1991) *For the Love of Enzymes: The Odyssey of a Biochemist.* Harvard University Press, p. 181.
28. AK Interview with Sally Smith Hughes, Program in the History of the Biosciences and Biotechnology, Biochemistry at Stanford, Biotechnology at DNAX, Arthur Kornberg, 1997, p. 31.
29. AK Interview with Sally Smith Hughes, Program in the History of the Biosciences and Biotechnology, Biochemistry at Stanford, Biotechnology at DNAX, Arthur Kornberg, 1997, p. 23.
30. Kornberg A. (1991) *For the Love of Enzymes: The Odyssey of a Biochemist.* Harvard University Press, p. 181.
31. Letter to Avram Goldstein, July 16, 1957. Reproduced with permission from the Department of Special Collections, Stanford University Libraries.
32. AK Interview with Sally Smith Hughes, Program in the History of the Biosciences and Biotechnology, Biochemistry at Stanford, Biotechnology at DNAX, Arthur Kornberg, 1997, p. 29.
33. AK Interview with Sally Smith Hughes, Program in the History of the Biosciences and Biotechnology, Biochemistry at Stanford, Biotechnology at DNAX, Arthur Kornberg, 1997, p. 30.
34. Kornberg A. (1991) *For the Love of Enzymes: The Odyssey of a Biochemist.* Harvard University Press, p. 235.
35. Kornberg A. (1991) *For the Love of Enzymes: The Odyssey of a Biochemist.* Harvard University Press, p. 236.
36. ECF Interview with Gerard Hurwitz, Oct. 2013.
37. Ibid.
38. Hurwitz J. (2005) The Discovery of RNA Polymerase. *J Biol Chem* 280: 41477–42485.
39. Kornberg A. (1991) *For the Love of Enzymes: The Odyssey of a Biochemist.* Harvard University Press, p. 236.
40. AK Interview with Sally Smith Hughes, Program in the History of the Biosciences and Biotechnology, Biochemistry at Stanford, Biotechnology at DNAX, Arthur Kornberg, 1997, p. 31.
41. Robert Baldwin, personal communication.

42. ECF Interview with Robert Baldwin, Aug. 2013.
43. AK Interview with Sally Smith Hughes, Program in the History of the Biosciences and Biotechnology, Biochemistry at Stanford, Biotechnology at DNAX, Arthur Kornberg, 1997, p. 29.
44. Letter from Arthur Kornberg to Robert Baldwin, May 27, 1959. Reproduced with permission from the Department of Special Collections, Stanford University Libraries.
45. ECF Interview with Robert Baldwin, Aug. 2013.
46. Letter from Arthur Kornberg to Joshua Lederberg, Aug. 19, 1958. Reproduced with permission from the Department of Special Collections, Stanford University Libraries.
47. Kornberg A. (1991) *For the Love of Enzymes: The Odyssey of a Biochemist.* Harvard University Press, p. 181.
48. Ibid.
49. AK Interview with Sally Smith Hughes, Program in the History of the Biosciences and Biotechnology, Biochemistry at Stanford, Biotechnology at DNAX, Arthur Kornberg, 1997, p. 41.
50. Friedberg EC. (2014) A Biography of Paul Berg, *The Recombinant DNA Controversy Revisited.* World Scientific Publishing Co., Singapore, p. 97.
51. AK Interview with Sally Smith Hughes, Program in the History of the Biosciences and Biotechnology, Biochemistry at Stanford, Biotechnology at DNAX, Arthur Kornberg, 1997, p. 3.
52. Letter to Henry Kaplan, Jan. 6, 1958. Reproduced with permission from the Department of Special Collections, Stanford University Libraries.
53. Letter from Gobind Khorana to Arthur Kornberg, June 17, 1957. Reproduced with permission from the Department of Special Collections, Stanford University Libraries.
54. Letter from Arthur Kornberg to Avram Goldstein, July 16, 1957. Reproduced with permission from the Department of Special Collections, Stanford University Libraries.
55. Letter from Arthur Kornberg to Henry Kaplan, Jan. 8, 1958. Reproduced with permission from the Department of Special Collections, Stanford University Libraries.
56. Kornberg A. (1991) *For the Love of Enzymes: The Odyssey of a Biochemist.* Harvard University Press, p. 181.
57. Ibid., p. 183.

58. DeCross Jacobs C. (2010) *Henry Kaplan and the Story of Hodgkin's Disease.* Stanford University Press, Stanford, Calif. p. 212.
59. Ibid., pp. 212, 213.
60. Ibid., p. 213.
61. Ibid., p. 220.
62. ECF Interview with Roger Kornberg, May, 2014.
63. DeCross Jacobs C. (2010) *Henry Kaplan and the Story of Hodgkin's Disease.* Stanford University Press, Stanford, Calif. p. 213.
64. Ibid., p. 220.
65. Ibid., p. 214.
66. Ibid.
67. Ibid.
68. Ibid.
69. Ibid., p. 215.
70. Ibid., p. 216.
71. Ibid., p. 217.
72. Ibid.
73. Ibid., p. 218.
74. Ibid., p. 221.
75. ECF Interview with Roger Kornberg, May 2014.
76. DeCross Jacobs C. (2010) *Henry Kaplan and the Story of Hodgkin's Disease.* Stanford University Press, Stanford, Calif. p. 221.
77. Ibid.
78. Ibid., p. 375.
79. Kornberg A. (1991) *For the Love of Enzymes: The Odyssey of a Biochemist.* Harvard University Press, p. 171.
80. Letter from Francis Crick to Arthur Kornberg, Oct. 12, 1959. Reproduced with permission from the Department of Special Collections, Stanford University Libraries.
81. AK Interview with Sally Smith Hughes, Program in the History of the Biosciences and Biotechnology, Biochemistry at Stanford, Biotechnology at DNAX, Arthur Kornberg, 1997, p. 91.
82. Ibid., p. 92.
83. Ibid.
84. Kornberg A. (1991) *For the Love of Enzymes: The Odyssey of a Biochemist.* Harvard University Press, p. 171.

85. Ibid., p. 172.
86. The Nobel Prize in Physiology or Medicine 1959. Severo Ochoa, Arthur Kornberg. http://www.nobelprize.org/nobel_prizes/medicine/laureates/1959/
87. Marianne Grunberg-Manago http://en.wikipedia.org/wiki/Marianne_Grunberg-Manago
88. Ibid.
89. Jerard Hurwitz, http://en.wikipedia.org/wiki/Jerard_Hurwitz
90. AK Interview with Sally Smith Hughes, Program in the History of the Biosciences and Biotechnology, Biochemistry at Stanford, Biotechnology at DNAX, Arthur Kornberg, 1997, p. 93.
91. Kornberg A. (1991) *For the Love of Enzymes: The Odyssey of a Biochemist.* Harvard University Press, p. 172.
92. Arthur Kornberg, Encyclopedia.com http://www.encyclopedia.com/topic/Arthur_Kornberg.aspx
93. Letter to California State Department of Education, Oct. 23, 1972. Reproduced with permission from the Department of Special Collections, Stanford University Libraries.
94. Document accompanying letter to Richard Dickerson, CalTech, 23 March, 1981. Reproduced with permission from the Department of Special Collections, Stanford University Libraries.
95. Letter to Cal Thomas, March 26, 1981. Reproduced with permission from the Department of Special Collections, Stanford University Libraries.
96. AK Interview with Sally Smith Hughes, Program in the History of the Biosciences and Biotechnology, Biochemistry at Stanford, Biotechnology at DNAX, Arthur Kornberg, 1997, p. 91.

CHAPTER EIGHT

The Stanford Department of Biochemistry

Kornberg thought long and hard about how best to organize his Stanford department, as well as settling it into the life of the medical school and the university. "His plan was to create the ideal work place, one in which individual research groups would cooperate with one other and nothing would interfere with the progress of their research endeavors," Buzz Baldwin wrote in a tribute to Kornberg shortly after his death in 2007.

> Rare enzymes would be shared and major instruments would be made available to everyone. Research grants would also be shared. Each faculty member was expected to acquire the funds spent by his (there were then no women faculty in the Department of Biochemistry) group, but strict accounting would not be required and there would be no financial deadlines. The entire department would be expected to attend noon seminars. The plan called for seminar presentations by faculty members, postdoctoral fellows, senior students and visiting scientists.[1]

A particularly innovative decision was to establish research laboratories occupied by graduate students and postdoctoral fellows from different departmental research groups, an arrangement that Kornberg initiated at Washington University that was designed to maximize the potential for inter-laboratory communication. "Arthur insisted that everyone in the department work together," Peter Lobban, a graduate student with Dale Kaiser, commented.

> You worked together; you shared reagents together; you talked to one another about science. We were all on a first name basis — except for some timid souls who referred to Arthur as Dr. Kornberg! So the notion of competition within the department did not exist. This provided a virtual intellectual paradise for working in.[2]

"Students and postdocs who have left have invariably said that this arrangement was one of the most enriching things they had had in their experience," Kornberg related.[3]

"When you share space, that means equipment, supplies, caretakers," Kornberg continued.

> The bookkeeping of who takes what, uses what, and so on would be impossible. So we shared all our resources. Some reagents were expensive, so there was no point in everybody having expensive radioactive reagents, especially early on when we were all doing similar experiments. We had a reputation of being well heeled, but that wasn't true. I as chairman, or my successors, would sometimes tell the faculty: "there's a moratorium; you can't buy any more equipment, and you have to go slow on supplies until the next funding period". Someone uses animals that are expensive; someone uses more expensive equipment. It was crucial that those of us who were in a positive balance would say: "Well, my colleagues are doing work similar to what I'm doing; we're sharing ideas and reagents and results. Okay, so right now I'm being more generous." That attitude prevailed. Young faculty members who came into the department were immediately full partners in the entire enterprise. They could take as many students and postdocs and reagents as anybody else, even though their funds were grossly inadequate for that. So there was a legacy of that kind of indebtedness and ultimately responsibility and sharing.[4]

Once ensconced in their new space at Stanford the faculty settled quickly into their research agendas, efforts considerably facilitated by the carefully planned logistics of the move from St. Louis. Melvin Cohn continued exploring his immunological interests. "When we got started at Stanford, the bulk of my research concerned the aminoacyl tRNA synthetases and their reaction with tRNAs," Paul Berg related. "There were also beginning attempts to isolate and characterize *E. coli* RNA polymerase

Paul Berg.

and efforts to obtain a robust cell-free system for protein synthesis."[5] In later years Berg's research contributions realized groundbreaking contributions to the development of recombinant DNA technology, work that earned him the Nobel Prize mentioned earlier. Now an Emeritus Professor at Stanford, Berg visits the office he occupies in the Beckman building on a daily basis.

Kornberg's early work on the enzymatic synthesis of DNA surfaced the imperative of characterizing the nucleases of *E. coli* that can degrade DNA, work that Bob Lehman initiated in St. Louis and continued soon after his arrival at Stanford. Lehman subsequently focused on the mechanism of genetic recombination, an effort that led to the discovery of the enzyme DNA ligase, as well as the recA recombinase. Dave Hogness and Dale Kaiser undertook studies aimed at transforming *E. coli* with DNA. Both went on to initiate important new research directions. "Dave's career at Stanford was brilliant — his work really founded the new field of developmental biology and he trained several of the modern leaders in this field," Buzz Baldwin related.[6]

Hogness made fundamental contributions to understanding the ontogeny of the fruit fly *Drosophila melanogaster*. He also investigated the role of the hormone ecdysone in the development of the fruit fly. In 1978,

Hogness and his group identified the TATA box (Goldberg–Hogness box) as the start sequence for the transcription of genes in eukaryotes.[7]

Dale Kaiser in turn initiated an innovative program in the study of swarming and fruiting body development in *Myxococcus xanthus*,

David Hogness.

Dale Kaiser.

a gram-negative, rod-shaped species of bacteria that exists as a predatory, saprophytic single-species biofilm called a swarm. At the time of this writing the 87-year-old Kaiser, now also an Emeritus Professor of Biochemistry at Stanford (as are Lehman, Hogness and Baldwin), continues to operate a research laboratory, and he too can be found in his office most days of the week.

Buzz Baldwin began work on DNA after his arrival at Stanford. "It is a measure of how interactive we were that Baldwin applied his skills and experience with physical chemistry to nucleic acids," Kornberg stated. "He made important contributions to the literature and to us."[8] In due course, all of the individuals just mentioned (with the exception of Mel Cohn) were elected to membership in the US Academy of Sciences.

Kornberg was highly respected by his faculty. In the tribute to Kornberg mentioned above, Baldwin noted that

> Arthur had a commanding presence; when he said something, people listened. I remember the story of Arthur speaking at a congressional hearing. After the hearing, one committee member was surprised to learn that another member had changed his vote and asked him why he did this. "I don't want to be called a fool by Arthur Kornberg" was the reply. [When] Dan Koshland, a longtime friend of Arthur's, was once speaking at an event that Arthur had organized he began by saying "I never say 'no' to Arthur Kornberg." On the other hand, Arthur would put you immediately at ease in a personal conversation. A friend once remarked, "he made you feel his whole attention was focused on you."[9]

"Arthur was also a great motivator," Paul Berg commented.

> He had this wonderful touch; being interested, warm — calling you in the middle of the night to find out if your experiment worked. He always stood for the right things. To him, honesty and integrity in science was uppermost. My thoughts about Arthur can be summarized thus: he could, in a conversation with you get you to agree to do something and make you feel that you'd come up with the idea yourself. He had an uncanny way of making you feel good about doing what he wanted you to do — and leaving you convinced that it was worth

doing.[10]

Kornberg ensured that all decisions, even important ones such as hiring new faculty members, were made democratically, though not necessarily by a strict majority vote. The idea was to have all opinions fully expressed and explored before he exercised a final decision in his capacity as chairman. "In 1963, we had an appointment that we wanted to make and we thought there was someone very attractive for the position," he related.

> We were six or seven at the time, and three of us thought that the individual was someone we wanted. The other three were uncertain. He turned out later to be an outstanding scientist, so we would have done very well to get him — but that's beside the point. Even though I had the authority as chairman and was in the group that wanted him very much, I deferred to those who had some doubts.

"It was crucial for organization of the department, and my being happy in it to have the department function in the spirit of a family rather than a department of the conventional kind," Kornberg related, a holdover from the way he had fervently cultivated his microbiology department at Washington University. "We were communal in our sharing of all resources — money, space, everything else."[11] "I have been fortunate in my family life — at home — and in the laboratory, too." Kornberg wrote. "More people have left than stayed. Those who stayed evolved the patterns and interests that gave this academic family its unique shape and character."[12]

The biochemistry department admitted only four new graduate students each year for a faculty of seven, and prospective students were carefully vetted. A good number became well-known investigators rising to the top of their fields. In 2013, one of his former trainees, Randy Schekman, was awarded a Nobel Prize, an event sadly missed by Kornberg, who was by then deceased.

"Research group sizes were kept small by national standards and faculty members frequently joined post-doctoral fellows and students at the laboratory bench," Buzz Baldwin wrote. "Students thus essentially learned the art and skills of laboratory work by the apprentice method."

Fundamentally, the department was very small relative to most biochemistry departments in the country and its unique shape and character, both in terms of faculty and staff and graduate students, allowed Kornberg to keep an eye on everything that transpired. "How did all this work in practice? Fabulously, according to postdoctoral fellows who, when they took up new jobs elsewhere wistfully recalled their days at Stanford," Baldwin related.[13]

****

Kornberg followed the progress of the research conducted by his own group with an eagle eye. In particular, he advocated a specific style by which he wished research notebooks be kept, as presently noted, an experience not universally enjoyed. And he insisted that these records always be available to him on the desks at which trainees sat. "Working with Arthur was not always easy, but it was always rewarding," Bob Fuller a former graduate student in the Kornberg group wrote in one of the many tributes to Kornberg published shortly after his death.[14] "When at an impasse in discussing results, or often the lack thereof, Arthur had the annoying habit of asking to see one's notebook and then critiquing its organization. I had many opportunities to have my notebook inspected, each successive time a bit more like Kabuki theater,"[15] Fuller wrote. But he also noted: "these encounters usually served to back me out of some experimental quicksand and steer me in new directions."[16]

Graduate student Lee Rowen worked in Kornberg's laboratory from 1972–1977 and had the singular distinction of being his first female graduate student. Rowen, whose passage through Kornberg's laboratory was less than plain sailing, opened an essay with the following comments:

> Everyone warned me. "Don't do it," they said. "You DO NOT want to work for Kornberg. You DO NOT have a strong enough personality." A.K. as we called him was not a warm and fuzzy advisor. His was more the tough love approach. At the time, it was acutely painful.[17]

"Arthur hated my notebook style," Rowen commented. "I used loose-leaf binders and occasionally a page would lack a title or date, or my decimal points wouldn't line up or, worst of all sins, I'd write with more than one color of ink. I got no end of complaints anytime I showed him data."[18]

"Arthur's lab was like the alchemical alembic," Rowen related.

> A.K. the master biochemist, transmuted coarse raw material — the sloppy, ill-informed, imprecise minds of graduate students — into the intellectual equivalent of gold. The process was difficult and troubling — but perhaps there is no other way to effect such wholesale transformation. I will always be grateful for what Arthur gave me. It was a privilege to be his student.[19]

Not everyone who passed through the Kornberg laboratory found his obsession with the way laboratory notebooks were kept offensive — or even unreasonable. Tania Baker, a former graduate student and postdoctoral fellow with Kornberg in the late 1980s is now a professor at The Massachusetts Institute of Technology. While acknowledging Kornberg's insistence on "a particular style that he wanted lab note books to be filled out in," Baker volunteered that

> that was the way his mind worked too. He would remember data that related to another piece of data 4 months back. For example if you had a graph that was related to another data graph back whenever and you had changed the color used, like for the control, versus the experimental line he'd become confused. Because he'd say — "but the red and the black were the other way around last time." — which was perhaps 4 months back!!! He had that kind of memory and that was the way his mind worked.

"When I was there we had a number of Japanese and other foreigners in the group, and some of the research projects overlapped each other quite a lot," Baker related.

> So when you presented group meeting you had to make a handout and you'd have to photocopy parts of your notebooks so that people could

follow what you were doing. And having everyone's lab notebooks in a sort of standard format was helpful in realizing that goal. So I didn't see it as personally intrusive. It seemed to be a practical thing to me.[20]

Paul Berg recalls no difficult experiences with Kornberg as a postdoctoral fellow in Kornberg's Washington University laboratory. When presented with a number of potential research projects for him to choose from, Berg outlined a project that he had independently identified based on work published by the European biochemists Fritz Lipmann and Feodor Lynen. After considerable discussion, Kornberg was ultimately persuaded about the cogency and appropriateness of the project. "'OK. Give it a shot,' he said. So that's what I started out to do. And he never looked at my notebooks or anything like that."[21]

"Every night, before he left the lab Arthur would stop by my bench and ask how things were going — and we'd sit and talk for a few minutes," Berg related. "When I eventually discovered that the results published by Lipmann and Lynen were in fact incorrect he was very pleased. Furthermore, when I wrote a paper on these studies Arthur did not add his name as a co-author."[22] "Arthur had a style that was different from mine and from everyone else in the Stanford Department of Biochemistry." Berg continued.

> My sense is that Arthur was from a generation of the great biochemists in the United States. He labeled himself as an enzymologist/biochemist and not as a molecular biologist. And he had little interest in genetics. He had the old style German geheimrat mentality. When he sat down with his graduate students, they had their notebooks with them and they went through these line by line. That was his style — he'd give them a bad time if their notebook was sloppy or if they couldn't explain something from their notes. He was of the opinion that this is the way graduate students should be taught.[23]

Roger Kornberg sheds a different light on Kornberg's single-minded attention to laboratory notebooks. "Arthur's influence was so pervasive and deeply affective not only because of his insistence on truth and rigor, but also because of his extraordinary attention to people," he commented.

"Who else has ever reviewed every notebook page of every student and fellow? Consider the time, effort and concern involved. He not only advised and taught by example, he was involved and personally engaged in the work of every student and fellow."[24]

"As much as Arthur was strict about his graduate students, he was an active and effective teacher," Berg stated.

> At Washington University we taught every day for two quarters. A huge amount of teaching. The course was for the medical students but it was open to anyone and the lecture hall was always packed. Arthur persuaded a lot of the clinical people to attend these lectures. He sat in on one's lectures and would offer friendly critiques after them — and he took his fair share of lectures.[25]

"Arthur was also very strict about students presenting seminars," Berg continued.

> In fact, I was once asked to give a Harvey Lecture and I gave it as a practice talk to the department. At the end of the talk Arthur came up

From left to right, Roger, Ken, Arthur and Tom Kornberg on the occasion of Roger receiving in the Nobel prize in 2006.

to me and said: "If you give that talk everyone will have either walked out — or be asleep. You are going to give a talk to a general audience. You have to make it simple." So he criticized a lot, but most of the time he was a very effective critic.[26]

The extent of Kornberg's passion and devotion to the efforts in his research groups over the years is perhaps most graphically illustrated by an anecdote from Bill Wickner, who as mentioned earlier was a postdoctoral fellow in Kornberg's group in the early 1970s. Struggling with a set of experiments, Wickner enjoyed the thrill of a significant advance late one Saturday night — more accurately early one Sunday morning. "At about 4:00 am I had the results confirmed," he related.

> I'm not the sort of guy who enjoys working in the lab at night. I like to sleep at night! But I was so excited I even failed to call my wife, who was probably wondering whether or not I'd been run over by a truck. And at four in the morning, in my miasma of exhaustion and excitement I decided to call Arthur. I found his home number and telephoned. He answered with a weary "Hello," and I said: "Arthur, it's Bill Wickner at the lab. I've got RNA-primed ribonucleoside triphosphate-dependent DNA synthesis to work and I knew you'd want to be the first to know." He said: "Stay right there," and hung up the phone. A voice in my head was screaming at me; a voice that began when I started dialing his phone number; a voice telling me to run from the lab, jump in the car, grab my wife and drive off to Alaska before he arrived. Arthur showed up in his bedroom slippers and bathrobe about ten minutes later and said: "Let's see the data." We spent a delightful hour going over the data, at the end of which he said: This is very exciting and you were right to call me at home. I'll be back at four o'clock this afternoon to see how things are progressing. Having woken me this morning I know you'll be continuing to work all day![27]

Arthur Kornberg's relationship with members of his laboratory is cogently articulated by Jack Griffith, one of the doyens of the electron microscopy of DNA, who spent the years between 1971–1978 as a postdoctoral fellow in the Kornberg Laboratory and subsequent years as a collaborator. "There was a great deal of awe about Arthur — in spite of his not infrequent critical mentality," Griffith commented.

I think it's fair to say that everybody who was around Arthur received the brunt of his criticism at one time or another. I certainly did! And I think it depended on how thick-skinned you were as to how you took that. I viewed it as serious fatherly criticism and tried to assimilate parts of Arthur's criticisms that I thought were well intended — and basically blew off the rest. Mainly, I felt: "Gosh here is someone who is so eminent, and if he has the time to think a little about how I am doing in science and is willing to spend time telling me what he thinks, that is certainly important to me, even if it wasn't quite what I wanted to hear all of the time."[28]

****

Well known to one another for many years, the initial biochemistry faculty group at Stanford related comfortably. However, tensions between Kornberg and Mel Cohn arose early, tensions that were seeded during the time that Cohn joined Kornberg's laboratory at Washington University. "Mel was working intensely on a basic question in immunology; whether one cell could make two different kinds of antibodies. With a catalytic mind and personality he was brilliant and engaging and intellectually involved," Kornberg related. "He was a very important person in St. Louis and could have been at Stanford as well. But there was a certain aloofness and maybe even contempt ------ that I found incompatible."[29]

"While at Washington University in St. Louis, Cohn was singularly disinterested and uninvolved in the preparations we were all making to move to Stanford," Kornberg continued.

When we got to Stanford there were many things to do of a custodial nature in getting settled. Again Mel was singularly disinterested. I think in a way he felt above it all. As the chairman I laid out the various responsibilities for the people in the group. On that list were some assignments to Mel, and he said: "You know I don't want to do that ----------- If I had to do all this crap I'd be chairman."

Kornberg's emphatic response was: "You're not chairman and this is what you need to do!" Shortly after the group's arrival at Stanford, Cohn informed Kornberg of a job offer he had recently had from Harvard. He was apparently shocked at Kornberg's response: "Well, maybe you ought to take it."[30]

"This was an example of the very direct way in which my father dealt with a lot of things — and that many people found off-putting," Roger Kornberg related. "He had legions of devoted former students. But at the same time there were those who disliked him intensely because he would not mince words. He told people exactly what he thought about any and all situations."[31]

Philosophical differences between Kornberg and Cohn did little to help forge a comfortable relationship between the two. Cohn "spoke enthusiastically about gestalt biology, the view that one cannot understand a biological problem by dissecting it into parts," Buzz Baldwin wrote. "Kornberg on the other hand naturally advocated using chemistry, specifically enzymes, as the basic tool for solving biological problems."[32] In 1962, Cohn left Stanford for the newly opened Salk Institute in La Jolla.

Following Baldwin's arrival and Cohn's subsequent departure, George Stark and Lubert Stryer joined the Department of Biochemistry in 1963, the second and third faculty appointments from outside the Washington University clique. Stark came on board as an Assistant Professor following postdoctoral training at Rockefeller University, having heard that Kornberg was eager to recruit a protein chemist. "Lubert Stryer and I were candidates at the same time and the biochemistry department liked us both," Stark commented.

> I was enormously impressed by Arthur's reputation and by his science once I understood what he was doing in his lab. At the time I was also enormously impressed by the way in which he ran the department. It was run very competently. And it was also run as a tremendously nurturing environment. And that was due to Arthur. The department was operated like a big club. We pooled our grant money; we had a common open stock room where everyone took whatever they needed. People

Arthur Kornberg in a relaxed mode — throwing a frisbee.

were mixed up in labs to encourage cross communication and cross-fertilization. So it was a truly unique environment and I've never seen anything like it elsewhere. And when Arthur stepped down it really didn't matter who the active chair was. The tradition was so ingrained. So that was Arthur and that side of him was remarkable. He had a reputation for driving the people in his lab very hard — the graduate students and post-docs. So people in his lab had a love-hate relationship with him. But no one outside his own group experienced that. He may have made critical comments when we presented work seminars — but that's all.[33]

Both Stark and Stryer were soon engaged in innovative experiments, especially developing new experimental methods that attracted wide attention. Stark, at the Lerner Research Institute of the Cleveland Clinic in Cleveland, Ohio, at the time of this writing was instrumental in detecting RNA molecules that had been separated by size by the use of electrophoresis, transferring them to chemically reactive paper and then using a hybridization probe complementary to part of or the entire target sequence, a technique that came to be dubbed Northern blotting (humorously based on the DNA blotting technique called Southern

Kornberg met with his students and postdoctoral fellows on a regular basis.

blotting, after its inventor Edward [Ed] Southern).[34] Stark and his colleagues are also credited with publishing the first paper that described the transfer of proteins by capillary action from a polyacrylamide/agarose gel onto a special membrane, a technique referred to as Western blotting.[35] Researchers have since referred to blotting of post-translational modifications, such as lipids and sugars, as Eastern blotting![36]

Stark's group also discovered N-phosphonacetyl-L-aspartate (PALA), an analog of aspartate transcarbamylase's transition state. They showed that PALA was a strong inhibitor of aspartate transcarbamylase and that it could enter mammalian cells to block pyrimidine nucleotide biosynthesis. With PALA, Stark and his colleagues went on to discover the giant polypeptide CAD that contains aspartate transcarbamylase, carbamyl phosphate synthetase and dihydro-orotase, all involved in pyrimidine synthesis. By studying CAD, Stark's group was one of the first to show gene amplification in mammalian cells.[37]

Kornberg had nothing but good things to say about George Stark, who he emphatically described as "a gem." In 1983, after 20 years at Stanford, Stark moved to the Imperial Cancer Research Fund in London and subsequently to the Cleveland Clinic. As for leaving Stanford, "I did so with great regret but with the firm conviction that it was time for a new environment," he related.

George Stark.

Lubert Stryer.

Part of the reason was the recognition that I did not want to spend my whole career in one place, part was a love for life in London where I had spent two sabbatical years, but the main part was a friendship and close collaboration with Ian Kerr, who had moved to the Imperial Cancer Research Fund from elsewhere in London. Together we helped discover the JAK-STAT pathway of signaling, which has been a mainstay of research in my lab ever since.[38]

Lubert Stryer and his coworkers pioneered the use of fluorescence spectroscopy, particularly Förster resonance energy transfer (FRET), as a major biophysical tool to monitor the structure and dynamics of biological macromolecules.[39] Stryer was also responsible for discovering the primary stage of amplification in visual excitation and showed that a single photoexcited rhodopsin molecule activates many molecules of transducin, which in turn activate molecules of a cyclic GMP phosphodiesterase. Stryer additionally contributed to understanding the role of calcium in visual recovery and adaptation.[40]

Kornberg described Stryer as "a very ambitious, brilliant, and mercurial person."[41] Red flags began waving when Kornberg vetoed a grant proposal that Stryer had put together on the grounds that he didn't have the space in which to use the extra money. Tensions escalated to the point that Kornberg was confronted by Stryer's demand to have an independent wing in the department assigned to him for his exclusive use. "Arthur was very supportive of Lubert," Bob Lehman recalled.

> But the department had been deliberately established with shared laboratories in order to promote cross-talk. Around the time that Lubert was developing FRET analysis he felt that he had to have his own dedicated space in order to attract the sort of individuals that he wanted in his laboratory. He threatened to leave the department if Kornberg did not accede to this request. Arthur said: "No way!"

"Despite our wish to adjust to individual faculty styles this issue was so crucial to the rest of us that we reluctantly let Stryer accept an offer from Yale that fulfilled his wishes for more autonomy," Lehman related.[42] Stryer left Stanford University for Yale in 1969 only to return to Stanford in 1976 as an endowed professor. Between 1975 and 1995 he was the sole author of four editions of a widely used textbook entitled *Biochemistry*. In 1995 he passed the baton to others, who produced additional editions of the book.

Another faculty appointment that didn't pan out well in the Department of Biochemistry was that held by James (Jim) Rothman, another intellectually outstanding young faculty member who was recruited in 1979 subsequent to Kornberg's retirement from the chairmanship of the department. "Jim is very smart — very bright," Paul Berg commented.

> But standoffish if he disdained one's intellectual level. He was extremely articulate at our Wednesday luncheon meetings — and also extremely critical when others spoke. He grew scientifically very rapidly. So his group size expanded. But he was not bringing in sufficient funds to support it financially. In addition, he had more space than anyone else. When discussions began concerning the new Beckman Center at the Medical School, Jim categorically told me that in his opinion the future of biochemistry and molecular biology was no longer in nucleic acids! The future was cell biology: understanding how cells get put together and how they work he said. He proposed that he would lead a group to move to the new center and the nucleic acid classical biochemistry groups should remain where they were. That was the sort of arrogance that characterized Jim Rothman.[43]

"There's no question that Jim is a brilliant scientist," Berg stated.

> He came to Stanford because of Arthur. He admired Arthur enormously while he was there. He wanted to study cellular components other than nucleic acids and he set up a beautiful system that allowed him to follow the movement of components from one cellular compartment to another. In fairness to him, he recognized that he was pushing the limits of what the department could provide. And that's why he ultimately left.[44]

"Jim always lived beyond his means," Kornberg stated. "He attracted many students and postdocs and expanded his base. Eventually it was clear that even though he knew and we knew that he was living beyond his means year after year, something had to give — and he left for Princeton."[45]

"Soon after Jim Rothman won the Nobel Prize he informed me that he always idolized my father and credits him for encouraging him to use enzymology in his own research," Roger Kornberg related. "But he also told me that when Arthur spoke to him directly his tone was typically nothing of the kind. The reality is that Jim was then so brash that Arthur once told him: 'Jim, you will destroy any department that you may join.'"[46]

The departures of Cohn, Stark, Stryer and Rothman were very much the exception. "Essentially all of us on the faculty had offers to go elsewhere at one time or another," Bob Lehman remarked. "But none of us took these seriously." Lehman was adamant about his own steadfast attachment to Kornberg and his department. "I loved Arthur," he stated.

> He was my scientific father — and a very good friend. For a period of about 10 years he and I lunched together on a daily basis. He was never devious. He was absolutely up-front with everyone. And he was relentless. Relentless is a good word to describe Arthur. He was relentless in his pursuit of scientific truth![47]

In 1971, Ron Davis joined the biochemistry department and never left Stanford. Davis brought with him the skill and art of DNA electron microscopy, a powerful tool that became widely used in the department. Douglas (Doug) Brutlag, a former graduate student in Kornberg's laboratory who subsequently cultivated a significant profile in bioinformatics was recruited to the faculty in 1974.

Beginning in 1959 the entire biochemistry department began convening at Kornberg's Portola Valley home to listen to the latest findings of a research group. "The whole department met in my living room once a month — and filled it with cigarette butts and so forth," Kornberg stated. "Those were the good old days when the group was small and speaking one language and infused with the excitement of discovery."[48] Within a few years the group could no longer fit comfortably into Kornberg's living room and the meetings were shifted to the Stanford campus. But the comfortable atmosphere that prevailed in Kornberg's home evaporated and in due course the get-togethers terminated. In 1972 the department initiated twice-a-year scientific retreats at Asilomar, a coastal meeting facility. Each such retreat, at which all the biochemistry research groups participated, lasted a few days.

Under Kornberg's leadership the Stanford Department of Biochemistry spawned a deliberately narrow research focus. "We didn't do research in carbohydrates, lipids, vitamins, minerals or bioenergetics," Kornberg related.

> 95% of what was considered important in a conventional biochemistry text was not pursued. Our focus on nucleic acids and proteins that interacted with nucleic acids was less than 5% of a biochemical textbook. But we were approaching this rather limited topic from the very broad standpoint of genetics, enzymology, and physical chemistry.[49]

****

The Stanford Dept. of Biochemistry — 1965–66.

Basic science departments in most American universities frequently (but by no means routinely) allow faculty in a particular department to hold a joint appointment in an academically related one, an arrangement that offers the potential for broadening the academic horizons of both concerned departments. When still mulling over the offer to leave Washington University for Stanford, Kornberg shared his interest with Henry Kaplan in recruiting Seymour Benzer, who as mentioned in an earlier chapter (see Chapter 7) was someone that Kornberg was very keen to bring to Stanford. A highly regarded geneticist and molecular biologist at Purdue University, Benzer was at one time Kornberg's choice to head the new Department of Genetics at Stanford. "I am sure we can provide a joint appointment in biochemistry for a person like Benzer or of comparable stature," he wrote to Kaplan.[50]

But once settled at Stanford, Kornberg evolved a contrary opinion about joint appointments involving the Department of Biochemistry, a mindset that led many at Stanford and at academic institutions in other parts of the country to speak critically of his attitude. Terms such as "arrogant" and "elitist" were not uncommon descriptors. Nor was Kornberg unaware of this reputation. "The Department of Biochemistry has had the reputation of being very exclusive, elitist," he told an interviewer.[51] Paul Berg conceded that.

> Kornberg's calculated philosophy of self-containment escalated to a reputation of aloofness among some members of the Stanford community — and beyond. "Snootiness" and "arrogance" were other descriptors that circulated. The reality is that the department didn't really interact with many outsiders.[52]

Considering the enormous efforts to which Kornberg went to generate the particular gestalt and organizational mentality that he established and cultivated in his small department, he had his own cogent reasons for refusing to allow joint faculty appointments. Most particularly he did not want "strangers" meddling with his organizational system in any way, shape or form. "We have not had the kind of joint appointments that are common in other institutions," he admitted.

> On the other hand, I've been willing to have someone called a professor of biochemistry and pediatrics so that he can maintain his identity with his Ph.D. in biochemistry. But I don't want him as an active member of the biochemistry department. So his official attachments are with another department. He can come to our seminars and we're eager to have him use our apparatus. Stanley Cohen, when he came to Stanford in 1968, used our centrifuges and our apparatus freely; all his early work with plasmids was done with our apparatus. He was really housed in this department. -------- But how big can a family be? How many people can you interact with, understand, be sympathetic to, be helpful to, get help from?[53]

Eric Shooter, a highly accomplished neurobiologist with special expertise in the biochemistry of brain proteins was the sole exception to Kornberg's policy concerning joint appointments. A protein chemist who obtained his PhD at Cambridge University in 1950, Shooter came to Stanford 10 years later to pursue a sabbatical in Buzz Baldwin's laboratory. "On the second day that I was in the department I went to visit with Arthur," Shooter related.[54]

> I had an older brother and we had both gone through Cambridge University. My brother was at the Chester Beatty Research Institute, concentrating on the physical chemistry of DNA. I, on the other hand, was a protein chemist. When I met Arthur, he looked at me quizzically and said: "I was expecting your brother!" I didn't have a reply to that so I simply kept quiet. Arthur asked: "You're going to work on DNA aren't you?" I told him that I was, with Buzz Baldwin.

Shooter and Baldwin carried out studies designed to determine whether DNA replication was indeed semi-conservative, as was then postulated, or conservative, an open question in the early 1960s.

At the end of his sabbatical year Shooter was offered a faculty position in the Department of Genetics at Stanford, where he carried out studies on the genetics of brain proteins and on nerve growth factor, studies that made important contributions to establishing the field of growth factors in the nervous system. Shooter's work gained him widespread attention and in due course he was offered the chairmanship of the Department of Neurobiology at the University of Oregon. The powers that be at Stanford worked hard to retain him, efforts that included an offer from Kornberg of a joint faculty appointment in the Department of Biochemistry. Nor was this single exception to Kornberg's otherwise inviolate policy concerning joint appointments a token one. Though he retained a laboratory in the Department of Genetics, Shooter participated fully in activities in the Department of Biochemistry, including teaching. In subsequent years, Shooter assumed the chairmanship of a new Department of Neurobiology at Stanford.[55]

In short, it is apt to quote the late Joshua Lederberg about the Stanford Department of Biochemistry under Kornberg's leadership:

> His talent and sensitivity as an administrator have built a Department of Biochemistry whose productivity is unmatched, and one which gives the lie to the proposition that science today is achievable only with immense groups and huge machines, or that it demands a renunciation of other human values.[56]

There is little question that Kornberg was imbued with what is archetypically referred to as a "strong personality." But truth be told, this descriptor translates to little more than the habit of thoroughly planning and implementing the manifold goals and ambitions that he embraced for much of his life, and executing them according to his own desires and dictates. He harbored firm opinions about issues that mattered to him and was unhesitating in expressing his views as situations dictated. Though unquestionably intimidating to some and possibly feared by others, he was widely liked and most certainly highly respected by those who were on more than polite greeting terms with him, most notably his faculty and colleagues in the international community of biochemists, as well as the great majority of the many postdoctoral and graduate students who over the course of many decades populated the various academic departments he led.

It can be fairly stated that Kornberg was not only widely respected, but revered. As pointed out by his son Roger, he was the chief apostle and leading practitioner of a new science that was predominant between 1945–1975, the science we now call classical biochemistry. He was a legend in his own time, a unique figure in the history of biochemistry, and a singular one in 20th century American science. His influence was pervasive and deeply affective, not only because of his insistence on scientific truth and rigor, but also as a direct consequence of his extraordinary attention to people. He not only advised and taught by example, he was involved and personally engaged in the work of every student and fellow in his laboratory. Such was his passion for the research he cultivated in his Stanford laboratory that Kornberg thought nothing of being awakened by telephone calls from his graduate students and postdoctoral fellows at all hours of the night to be informed of the outcome of exciting experiments. He also played seminal roles in the growth and maturation of the Stanford University School of Medicine as a whole and in the advancement of the Stanford Department of Chemistry.[57]

As an unadulterated biochemist with a particular passion for enzymology, Kornberg harbored no special interest in exploring the genetics or biology that may lie behind his biochemical wizardry, though he certainly valued the efforts of others that did. Nor was he unappreciative of the stunning progress in the discipline of molecular biology which blossomed during the latter part of the 20th century. But above all he firmly believed that prokaryotes, especially the well-characterized bacterium *Escherichia coli*, were model organisms with indispensable utility for understanding the majority if not all aspects of the biology of higher organisms. The adage "what is true for *E. coli* is true for the elephant" is one that Kornberg took to heart.

Having begun their academic association with Kornberg as postdoctoral fellows when he was at Washington University, most of his initial faculty at Stanford University shared this view early in their research careers. However, some eventually adopted new research aspirations for which they selected different model organisms, a strategy that Kornberg sometimes took issue with. When in the late 1960s, Paul Berg informed Kornberg that he had elected to abandon the world of microorganisms to study gene function in higher organisms, Kornberg openly expressed shock and dismay, convinced that Berg was making a huge career mistake in deserting the bacterium *E. coli* as a model organism. Berg in turn challenged his mentor about what he considered to be Kornberg's narrow vision about his (Berg's) decision to abandon studies on prokaryotes to understand aspects of gene function in higher organisms. "Arthur sometimes talked about 'fashions in research' and lamented the contemporary vogue of discarding simple prokaryotic systems that had been so informative over the past 30 to 40 years," Berg related. Others shared this view of Kornberg's passionate devotion to prokaryotes as model organisms. "Arthur was not especially pleased when Paul Berg, Dale Kaiser and David Hogness branched out into other areas," Eric Shooter remarked. "That was not really to his liking. He wanted everyone to work on nucleic acid biochemistry in *E. coli*."[58]

Arthur's eldest son Roger emphatically disagrees with the opinion that his father harbored such narrow views. "He not only appreciated the importance of studies in eukaryotic systems, but at one point he actually worked briefly in that area," Roger related. "Indeed, during the summer between high school and college, I worked with him in his laboratory on histones and chromatin. And he was a strong supporter of my pursuit of chromatin structure and later on of my studies of eukaryotic transcription in yeast."[59]

Such philosophic "clashes" and the emotions they may have sometimes provoked were infrequent and short-lived. Furthermore, the colleagues most likely to surface Kornberg's ire were typically not among his faculty or colleagues, but the graduate students and postdoctoral fellows in his own research group, from whom he not only expected uninterrupted and productive research, but also a style and manner of working in the laboratory — in particular, maintaining their laboratory notebooks in a manner of his choosing.

References

1. Baldwin R. (2008) Recollections of Arthur Kornberg (1918–2007) and the Beginning of the Stanford Biochemistry Department. *Protein Science* 17: 385–388.

2. Friedberg EC. (2014) *A Biography of Paul Berg — The Recombinant DNA Controversy Revisited*. World Scientific Publishing Co., Singapore, p. 162.
3. AK Interview with Sally Smith Hughes, Program in the History of the Biosciences and Biotechnology, Biochemistry at Stanford, Biotechnology at DNAX, Arthur Kornberg, 1997, p. 34.
4. Ibid.
5. Berg P, personal communication, Feb. 2015
6. Baldwin R, personal communication, Dec. 2014.
7. David Hogness, http://en.wikipedia.org/wiki/David_Hogness
8. AK interview with Sally Smith Hughes, Program in the History of the Biosciences and Biotechnology, Biochemistry at Stanford, Biotechnology at DNAX, Arthur Kornberg, 1997, p. 43.
9. Baldwin R. (2008) Recollections of Arthur Kornberg (1918–2007) and the Beginning of the Stanford Biochemistry Department. *Protein Science* 17: 385–388.
10. ECF Interview with Paul Berg, Nov. 2013.
11. AK Interview with Sally Smith Hughes, Program in the History of the Biosciences and Biotechnology, Biochemistry at Stanford, Biotechnology at DNAX, Arthur Kornberg, 1997, p. 34.
12. Kornberg A. (1991) *For the Love of Enzymes: The Odyssey of a Biochemist*. Harvard University Press, p. 177.
13. Baldwin R. (2008) Recollections of Arthur Kornberg (1918–2007) and the Beginning of the Stanford Biochemistry Department. *Protein Science* 17: 385–388.
14. Fuller RS. (2008) A Tribute to Arthur Kornberg 1918–2007. *Nature Struct & Mol Biol* 15: 2–17.
15. Ibid.
16. Ibid.
17. Ibid.
18. Rowen L. (2008) Tribute to Arthur Kornberg, Stanford University School of Medicine, January 25, 2008, pp. 79–80 (private publication).
19. Ibid., pp. 79–81.
20 ECF interview with Tania Baker, Nov. 2013.
21. ECF Interview with Paul Berg, Nov. 2013.
22. Ibid.
23. Ibid.
24. Roger Kornberg, personal communication, Aug. 2014.
25. ECF Interview with Paul Berg, Nov. 2013.

26. Ibid.
27. ECF Interview with William Wickner, Nov. 2014.
28. ECF Interview with Jack Griffith, Dec. 2014.
29. AK Interview with Sally Smith Hughes, Program in the History of the Biosciences and Biotechnology, Biochemistry at Stonford, Biotechnology at DNAX, Arthur Kornberg, 1997, p. 32.
30. Ibid.
31. ECF Interview with Roger Kornberg, May 2014.
32. Baldwin R. (2008) Recollections of Arthur Kornberg (1918–2007) and the Beginning of the Stanford Biochemistry Department. *Protein Science* 17: 385–388.
33. ECF Interview with George Stark, Dec. 2013.
34. Northern Blot, http://en.wikipedia.org/wiki/Northern_blot
35. Mukhopadhyay R. (2012) The Men Behind Western Blotting. *ASBMB Today*, March 2012. [http://www.asbmb.org/asbmbtoday/asbmbtoday_article.aspx?id=16084]
36. Eastern blot. http://en.wikipedia.org/wiki/Eastern_blot
37. Mukhopadhyay R. (2012) Stark Raving Mad For Science, *ASBMB Today*, June 2012. [http://www.asbmb.org/asbmbtoday/asbmbtoday_article.aspx?id=17032]
38. Stark G, personal communication, March 2015.
39. Forster Resonance Energy Transfer, http://en.wikipedia.org/wiki/F%C3%B6rster_resonance_energy_transfer
40. Lubert Stryer, http://en.wikipedia.org/wiki/Lubert_Stryer
41. AK Interview with Sally Smith Hughes, Program in the History of the Biosciences and Biotechnology, Biochemistry at Stanford, Biotechnology at DNAX, Arthur Kornberg, 1997, p.35.
42. ECF Interview with Robert Lehman, Oct. 2014.
43. ECF Interview with Paul Berg, April 2014.
44. Ibid.
45. AK Interview with Sally Smith Hughes, Program in the History of the Biosciences and Biotechnology, Biochemistry at Stanford, Biotechnology at DNAX, Arthur Kornberg, 1997, p. 35.
46. ECF Interview with Roger Kornberg, Jan. 2014.
47. ECF Interview with Robert Lehman, Oct. 2014.

48. AK Interview with Sally Smith Hughes, Program in the History of the Biosciences and Biotechnology, Biochemistry at Stanford, Biotechnology at DNAX, Arthur Kornberg, 1997, p. 43.
49. Ibid., p. 42.
50. Letter from Arthur Kornberg to Henry Kaplan, June 6, 1958. Reproduced with permission from the Department of Special Collections, Stanford University Libraries.
51. AK Interview with Sally Smith Hughes, Program in the History of the Biosciences and Biotechnology, Biochemistry at Stanford, Biotechnology at DNAX, Arthur Kornberg, 1997, p. 25.
52. Friedberg EC. (2014) *A Biography of Paul Berg — The Recombinant DNA Controversy Revisited*, World Scientific Publishing Co., Singapore, p. 97.
53. AK Interview with Sally Smith Hughes, Program in the History of the Biosciences and Biotechnology, Biochemistry at Stanford, Biotechnology at DNAX, Arthur Kornberg, 1997, pp. 42, 43.
54. ECF Interview with Eric Shooter, Oct. 2014.
55. Ibid.
56. Lederberg J. (1991) Foreword, Kornberg A. *For the Love of Enzymes: The Odyssey of a Biochemist*. Harvard University Press, p. ix.
57. Roger Kornberg, personal communication, May 2014.
58. ECF Interview with Eric Shooter, Oct. 2014.
59. Roger Kornberg, personal communication, May 2014.

CHAPTER NINE

Life in the Test Tube?

Following the many invited seminars and lectures that Arthur Kornberg presented during the 10 years between 1957 and 1967, he was frequently asked the vexing question: "Why have you been unable to replicate the genetic activity of the DNA product synthesized *in vitro*?" "Of course, we had tried many times to replicate the DNAs from *Pneumococcus*, *Hemophilus*, and *Bacillus subtilis*, each with readily measured transforming activity," he wrote. "These DNAs performed adequately as template-primers for net synthesis of DNA, but there was always a net loss rather than an increase of biological activity."[1]

The solution to this conundrum ultimately surfaced when Kornberg switched his focus to a bacterial virus called ɸX174 (which bears the distinction of being the first DNA-based genome to be sequenced), which he obtained from Robert (Bob) Sinsheimer, then at the University of Iowa, as part of a collaborative research effort. The single-stranded genome of this virus is comprised of a mere 5000 nucleotides; about five gene equivalents, and the notable lack of free ends suggested that it was almost certainly circular — or nearly circular. Within seconds after the viral DNA enters *E. coli* DNA polymerase converts it from a single stranded to a conventional duplex state. It was additionally shown that naked ɸX174 DNA could by itself infect a cell. Without its protein coat, which the virus uses to attach itself to the surface of the cell and which makes every virus particle infectious, uptake of naked viral DNA by a cell is a rare event. "Nevertheless, this infectivity by free DNA could be measured with confidence and accuracy," Kornberg related.[2] When synthesis was vastly extended the composition and the frequency of the nearest neighbor sequences indicated that both the φX174 DNA and the complementary strand product served as templates; each matched by synthetic strands oriented in the opposite direction. This double-stranded DNA product directed further replication as effectively as common DNA templates.

"The barrier that stood in the way of our synthetic product becoming infectious was the requirement that it be circular," Kornberg wrote.[3] Unlike the DNA from the virus, the Kornberg product was a linear chain and would have to have its ends joined to be capable of infecting *E. coli*. "Many of us believed that an enzyme must exist in *E. coli* that can ligate properly abutted ends of a DNA chain to generate a circle, as well as seal breaks in DNA that are intrinsic to manipulating any DNA in the test tube," Kornberg surmised.[4] When in 1961 five research groups independently discovered this enzyme (appropriately named DNA ligase) — those of Marty Gellert at the NIH, Charles Richardson at Harvard, Bob Lehman at Stanford, Jerry Hurwitz in New York and Kornberg with postdoctoral fellow the late Nick Cozzarelli — the goal of generating circular φX174 DNA was within reach. Kornberg wrote:

> With ligase in hand to circularize a linear DNA and with the Sinsheimer group eager to assay the infectivity of our products the stage was set for me and Mehran (Mickey) Goulian, my post-doctoral fellow, to determine whether our DNA polymerase could make biologically active DNA from the four nucleotides.[5]

To their intense satisfaction the pair were able to replicate the single-stranded circle of phage φX174 with DNA polymerase and then seal the complementary product with DNA ligase. The circular product was isolated and replicated to produce a circular copy of the original viral strand, which could be assayed for infectivity in *E. coli*. Kornberg and Goulian observed the synthetic viral strand to be as infectious as that of the phage DNA with which they started. The excitement was immense! "After twelve years of trying, we had finally done it — we had gotten DNA polymerase to assemble a nucleotide DNA chain with the identical form, composition, and genetic activity of DNA from a natural virus," Kornberg wrote.

> All the enzyme needed was the four common building blocks: A, G, T, and C. At that moment, it seemed there were no major impediments to the synthesis of DNA, genes, and chromosomes. The way was open to create novel DNA and genes by manipulating the building blocks and their templates.[6]

In writing about this watershed experiment in his autobiography, Kornberg was moved to lofty comparisons.

> In a very small way, we were observers to something akin to what those at Alomogordo on a July day in 1945, witnessed in the explosive force of the atomic nucleus. Harnessing the enzymatic powers of the cellular nucleus had neither the dramatic staging of light and sound of the atomic bomb nor the stunningly apparent global consequences. Yet, this demonstration of our power with enzymes that build and link DNA chains would soon help others forge a different revolution, the engineering of genes and the modification of species.[7]

When reporting the results of these experiments in the literature Kornberg cogently pointed out that while

> the fact that *E. coli* DNA polymerase can synthesize biologically active DNA does not establish its function in the replication of the bacterial chromosome. However, the effectiveness of the combined action of the polymerase and the polynucleotide-joining enzyme in forming infective DNA may have considerable significance for chromosomal replication.[8]

A later chapter reveals that subsequent studies by others demonstrated the existence of multiple DNA polymerases in *E. coli* and that while the so-called Kornberg DNA polymerase is indeed required for DNA replication in prokaryotes, it is not the enzyme that copies the template DNA.

These cautionary caveats notwithstanding, there is little question surrounding the significance of the experiments reported by Goulian,

Kornberg and their collaborator Bob Sinsheimer, which surfaced in the scientific literature as paper XXIV in the series *Enzymatic Synthesis of DNA*, with the triumphant subtitle *Synthesis of infectious phage φX174 DNA*. But as Kornberg disdainfully pointed out: "to the editors who sent the newsmen, it seemed that a virus had been synthesized and life created in the test tube."[9]

The media swarmed the Stanford campus in droves. On December 14, 1967 a hundred newspaper and television reporters and photographers attended a press conference called by Spyros Andreopoulos, Director of the Stanford News Bureau. Kornberg was relentlessly besieged with the question: "Is this DNA you've made a 'living molecule?'" "At the news conference I tried hard to explain why the answer to that question was so elusive," he pleaded. "Viral DNA has no life of its own, nor does the virus that bears it in the sense of being able to grow and reproduce outside the cells of an organism."[10] He evidently succeeded in popping this lead balloon to some extent, since when the Stanford conference concluded he overheard a reporter on the telephone to his office state: "It's not what we expected. They haven't made a virus. It's only a molecule, a short chain of DNA. They've been making DNA in the test tube for 12 years!" Yet, the story rated banner headlines worldwide and a newspaper article on January 4, 1968, was entitled: "Creation of Life Rates Best of Science Stories in 1967." In smaller type: "Human Heart Transplant Second."[11]

To add further fuel to the now somewhat embarrassing flames leaping off television screens and news pages, President Lyndon Johnson read a prepared televised statement during a speech at the Smithsonian Institute, which was then coincidentally celebrating the bicentennial of the *Encyclopedia Britannica*. That evening the Kornberg/Goulian experiments were the lead story on the televised news, which featured the President extemporaneously stating: "Some geniuses at Stanford University have created life in the test tube!"[12] In keeping with his long held advocacy of financial support for basic bio-medical science (see Chapter 15), Kornberg's measured response to Johnson's statement was that "ironically, Johnson had already begun the deceleration of government support of basic research that was to set a pattern for subsequent administrations."[13]

Mehran Goulian.

Years later Kornberg modestly told interviewer Sally Hughes:

> The popular response to the experiments [that Goulian and I reported] was excessive. It was over appreciated in terms of what it actually represented. It was something like Dolly.* It was on the front pages that we had created life in the text tube. As I mentioned in my book [*For the Love of Enzymes: The Odyssey of a Biochemist*] Max Perutz, the venerable scientist in Cambridge, England, wrote a rather caustic letter to the *London Times* saying, in effect, "What is all the hoopla about? Kornberg's doing things that were scientifically anticipated. It isn't all that novel." I agreed with him about the excessive publicity, but like him, I failed to anticipate that this discovery and related ones to follow would rapidly lead to the emergence of recombinant DNA technology and the extraordinary applications of genetic engineering.[14]

Kornberg also cogently pointed out that the fact that he and Goulian had utilized synthetic nucleotides in their experiments proved that infectious DNA did not contain novel nucleotides. "That had never been known before. It's as simple as A's, T's, G,s and C's," he noted. But importantly, the pair stressed that this reality opened the door to

---

* Dolly, an ewe, was the first mammal to have been successfully cloned from an adult cell. She was cloned at the Roslin Institute in Midlothian, Scotland, and lived there until her death when she was six years old.

site-directed mutagenesis by introducing a novel nucleotide that would be the source of a mutation.[15]

****

From the very get-go, all *E. coli* DNA polymerase fractions manifested a persistent 3' exonuclease activity, regardless of the extent of purification. Kornberg first adhered to the most obvious explanation: The nuclease was a contaminant that was difficult to get rid of by the enzyme purification schemes he and his collaborators used. Buzz Baldwin disagreed. "I'll bet you a bottle of champagne that the nuclease activity in DNA polymerase is part of the enzyme," he bravely announced.[16] But it made no sense to Kornberg that an enzyme that synthesized DNA would also degrade that substrate. So he happily took Baldwin's bet. It cost him a bottle of champagne!

In early 1964 Bob Lehman, together with postdoctoral fellow Charles Richardson, demonstrated that the exonuclease function of *E. coli* DNA polymerase exclusively attacks DNA starting at the 3'OH end of a DNA chain, yielding deoxynucleoside 5'-phosphates as products. "It is closely associated physically with DNA polymerase and has as yet not been dissociated from it," Lehman and Richardson noted.[17] A decade later Doug Brutlag demonstrated that the exonuclease activity was far more potent on single-stranded DNA than on the double-stranded form. In a series of elegant experiments Brutlag demonstrated that the 3' exonuclease function of DNA polymerase I removes incorrect nucleotides incorporated during DNA synthesis. Brutlag and Kornberg thus correctly postulated that the exonuclease activity is indeed an integral component of *E. coli* DNA polymerase and that it subserves a proofreading function by removing misincorporated nucleotides during DNA synthesis.[18]

Further studies by Murray Deutscher in the Kornberg laboratory revealed that *E. coli* DNA polymerase is endowed with a second exonuclease — a physically inseparable activity that removes mispaired nucleotides from the 5'-end of a polynucleotide chain. This 5' exonuclease was shown to also possess a phosphatase activity, facilitating 5'→3' exonucleolytic activity that removes either –OH or –P termini. Kornberg and his colleagues reasoned that the 5′ exonuclease function of *E. coli* DNA polymerase allows for a process called nick translation. DNA isolated in the laboratory is far from structurally pristine. Among other possible types of "deformities" DNA molecules that have been handled in the test tube typically contain multiple random single strand breaks, or "nicks," as they as colloquially referred to. These can be visualized by electron microscopy by dint of DNA polymerase binding at these sites. DNA synthesis cannot initiate from nicks, but removal of nucleotides by the 5′→3′ exonuclease function of DNA polymerase was persuasively postulated to allow filling of the resulting gap in the affected strand by DNA synthesis across the gap, using the opposite intact strand as a copying template. At this point replication ceases, leaving a new nick further downstream — hence the moniker *nick translation.*

Kornberg asked the rhetorical question: "Is there any point to this apparently futile exercise by DNA polymerase in removing DNA from the 5′ end, thus creating a gap, while filling it by extending the 3′ end?" Years later a critical role for such an activity for the initiation of DNA synthesis in living cells was identified (see Chapter 12). But in the late 1960s the most reasonable explanation offered was that nick translation might be fundamental to excising short segments of DNA from a strand containing sites of base damage that can arrest DNA replication, i.e., repairing DNA by excising short segments from a strand burdened with nucleotides structurally altered by exposure to one or more exogenous agents — such as ultraviolet (UV) radiation, an explanation fully consistent with earlier studies by investigators primarily interested in the molecular mechanisms by which bacteria repair damaged DNA.[19] Indeed, beginning in the mid-1960s, researchers in the DNA repair community had studied the events that transpire in DNA containing pyrimidine dimers, photoproducts in which two adjacent pyrimidines are covalently joined following exposure of DNA to ultraviolet (UV) radiation. Accordingly, it was known that bacteria such as *E. coli* are endowed with a specific mechanism for excising pyrimidine dimers as well as other types of base damage that perturb the normal helical structure of the genome. The excision of pyrimidine dimers in DNA generated by exposure to UV light is an especially well-studied example of what is now referred to as *nucleotide excision repair* (NER). Though the precise mechanism of NER was then far from understood it was widely assumed that damaged DNA is *incised* (nicked) with defined polarity close to a lesion such a pyrimidine dimer by a specific endonuclease, facilitating *excision* of the lesion by a second nuclease function. This process necessarily leaves gaps in the damaged strand that were believed by DNA repairologists to be filled in by one or more DNA polymerases, a belief that in due course became established fact.[20]

In late 1969 Kornberg and his colleagues published observations documenting the excision of thymine-thymine dimers by the 5′→3′ exonuclease function of DNA polymerase *in vitro*, observations that they agreed were fully consistent with a DNA repair function. They wrote:

> The multiple catalytic activities of purified DNA polymerase from *E. coli*, permitting polymerization and hydrolysis have been recognized previously, but the biological significance of the enzyme's potentiality for nick translation in duplex DNA has been obscure. It is now evident that the versatility of the 5′→3′ exonuclease activity is greater than had been recognized, and that its ability to excise mismatched sequences by hydrolyzing phosphodiester bonds in the hydrogen-bonded region on the 3′ side of pyrimidine dimers or other distortions in the polynucleotide permits DNA polymerase to play a role in DNA repair (and in recombination) in addition to its role in replication.[21]

Kornberg was sufficiently excited by this observation that in July 1962 he wrote to Francis Crick informing him:

> DNA polymerase does a dandy job in excising pyrimidine dimer lesions by its 5′→3′ exonuclease function. The enzyme, by its replicative nick translation coupled with "endonucleolytic" excision of mismatched sequences, may be a candidate for the *in vivo* job.[22]

Hence, the obvious question arose: did *E. coli* DNA polymerase evolve specifically to participate in excision repair of DNA or to synthesize DNA during semi-conservative DNA replication — or conceivably both? This question clearly begged studies that addressed the phenotype of mutants defective in this enzymatic function. To the best of the author's knowledge Kornberg made no attempt to isolate a mutant *E. coli* strain defective in the DNA polymerase activity he discovered. But even if he had contemplated such an undertaking it was not at all obvious what sort of genetic selection or screen might lend itself to the task of efficiently screening billions of mutated cells. But John Cairns cheerfully assumed this formidable logistical challenge.

An Australian by birth, Cairns acquired an MD degree from Oxford University in 1946. He subsequently worked as a virologist at the Walter and Eliza Hall Institute of Medical Research in Melbourne, Australia and at the Virus Research Institute at Entebbe, Uganda, before returning to Australia to join the School of Microbiology at the John Curtin School of Medical Research. Beginning in 1960 Cairns spent two years at the Cold Spring Harbor Laboratory and served as director of the Laboratory from 1963 to 1968. He remained a staff member at Cold Spring Harbor until 1972, at which time he was appointed head of the Mill Hill Laboratory of the Imperial Cancer Research Fund in the UK. After leaving Mill Hill in 1980 he took up a professorship at the Harvard School of Public Health. Cairns retired in 1991.[23]

John Cairns.

Undaunted by the Herculean task of painstakingly examining one bacterial colony at a time, Cairns and his research colleague Paula deLucia undertook the arduous labor of searching for a mutant strain of *E. coli* defective in DNA polymerase activity. The pair systematically examined extracts of individual bacterial colonies exposed to a mutagen one by one, until after examining over 3,000 colonies they identified the quarry they were hunting. In confirmation of the notion that these results strongly supported a DNA repair function for the Kornberg DNA polymerase, the mutant was subsequently shown to be abnormally sensitive to killing following exposure to UV light. An obvious implication of these observations was that the enzyme purified and characterized by Kornberg and his colleagues was not by itself directly responsible for semi-conservative DNA replication, if at all, strongly suggesting that *E. coli* and presumably all other cells in nature are endowed with more than one DNA polymerase.

Rick Horwitz from the University of Virginia recollects being in Kornberg's office when John Cairns called to inform Kornberg about the viable DNA polymerase-defective mutant that he had isolated.

> There had been a rumor floating around Stanford that Cairns had such a mutant and that the cells were "OK," Horwitz related. [Arthur] and I were doing one of our one-on-one magnetic resonance tutorials and the phone rang in his office. I asked him if I should leave and he said no, you can stay. After the call, he said that he had received an interesting call from John Cairns and then told me about the mutant. I asked him what he was going to do and he said that they had looked hard for other [DNA] polymerases and hadn't seen any — and now this opened up a great opportunity — and he seemed genuinely excited about it.[24]

Suggestions that the DNA polymerase isolated and characterized by Kornberg and his colleagues was in fact not the replicative polymerase surfaced in editorials in *Nature* published in 1969 and 1970. The 1969 *Nature* editorial entitled *What Makes DNA Duplicate?* opined:

> ---- there have always been objections to the idea that the Kornberg enzyme was responsible for semi-conservative DNA replication in *E. coli*. For one thing, the Kornberg enzyme fails to duplicate double stranded

DNA *in vitro*, and for another, mutants of *E. coli* have been isolated which are defective in DNA duplication but have normal Kornberg polymerase activity.[25]

The 1970 *Nature* editorial, dubbed "Multifarious Molecules," addressed studies on DNA polymerase by two Danish investigators, Hans Klenow and Jørgen Overgaard-Hansen. "Klenow and Overgarrd-Hansen have treated the enzyme with a protease, subtilisin," the editorial read.

> When DNA is absent, this causes loss of exonuclease, but not of polymerase activity. When DNA is present, both activities remain substantially intact, but fractionation by gel filtration reveals the presence of two fragments, one possessed only of polymerase, the other only of nuclease activity. ---------- The inference is that DNA polymerase consists of two parts, joined perhaps by only a single exposed loop of polypeptide chain that can be cleaved without other damage to either half of the molecule. This evidently is a case in which two different functional subunits are tethered together.[26]

In subsequent years the exonuclease fragment of DNA polymerase that retains the polymerization and $3' \rightarrow 5'$ exonuclease activity became colloquially referred to as the "Klenow fragment."

Kornberg was offended by the *Nature* editorials. During a sabbatical in 1970 he was in residence at the Laboratory of Molecular Biology in Cambridge, UK sharing an office with Francis Crick and Sydney Brenner, and at one point he communicated his displeasure about the comments concerning DNA polymerase. Crick, who in turn had been agitated almost to the point of fury that reverse transcriptase, an enzyme that copies RNA to yield DNA, had violated his central dogma of DNA to RNA to proteins, was sympathetic. "Forget it," was Crick's sober advice to Kornberg. "It's like trypsin and chymotrypsin; they have essentially the same function but they are designed a little differently. When they find [a] new [DNA] polymerase, it will be much like what you've described."[27]

In his autobiography Kornberg facetiously wrote:

> "DNA polymerase, you are charged with masquerading as a replication enzyme. Your ability in repairing DNA damage has been misrepresented

by your agents as relevant to replication. You are a red herring." Such, in paraphrase, were the accusations against DNA polymerase and me by *Nature New Biology* in a series of unsigned, defamatory editorials.[28]

Clearly, understanding the precise biological function(s) of the *E. coli* DNA polymerase discovered by Kornberg was becoming complicated.

By now other laboratories were honing in on the conundrum. In May 1970 Friedrich Bonhoeffer at the Max Planck Institute in Germany announced in *Nature* an *in vitro* system that supported DNA synthesis, which was robust in the Cairns mutant.[29] The same issue of *Nature* contained an article by Rolf Knippers and Wolf Stratling that reported DNA synthesizing activity in the Cairns mutant.[30] In December 1970 Knippers reported the detection of a novel membrane-associated DNA polymerase activity in the Cairns mutant that he dubbed DNA polymerase II,[31] the Kornberg enzyme being now identified as DNA polymerase I.

But Kornberg was not ready to enter a plea of guilty. Alternative explanations for the observations on John Cairns' mutant were theoretically viable — and he suggested several. "Possibly the low activity of polymerase in the Cairns mutant might be due to poor extractability of the enzyme rather than its absence in the intact cell," he wrote. "Perhaps the enzyme, though defective, could still function inside the cell but not in the artificial environment of an extract. Conceivably, the normal abundance might be so generous that a residual one percent suffices to sustain replication."[32]

As recounted more fully in the next chapter, intensely satisfying to him professionally — and personally and in almost improbable fairy-tale fashion, significant clarification to this conundrum came from none other than one of Arthur Kornberg's sons, Tom, then an accomplished cellist of concert caliber in training at the famed Julliard School in New York and coincidentally simultaneously a graduate student in molecular biology at Columbia University.

References

1. Kornberg A. (1991) *For the Love of Enzymes: The Odyssey of a Biochemist.* Harvard University Press, p. 192.

2. Ibid., p. 194.
3. Ibid., p. 196.
4. Ibid.
5. Ibid.
6. Ibid., p. 199.
7. Ibid., p. 200.
8. Goulian M, Kornberg A, Sinsheimer RL. (1967) Enzymatic Synthesis of DNA, XXIV. Synthesis of Infectious Phage oX174 DNA. *Proc Natl Acad Sci* **58**: 2321–2328.
9. Kornberg A. (1991) *For the Love of Enzymes: The Odyssey of a Biochemist.* Harvard University Press, p. 201.
10. Ibid.
11. Ibid., p. 206.
12. Ibid., p. 203.
13. Ibid.
14. Ibid., p. 206.
15. AK Interview with Sally Smith Hughes, Program in the History of the Biosciences and Biotechnology, Biochemistry at Stanford, Biotechnology at DNAX, Arthur Kornberg, 1997, p. 68.
16. Kornberg A. (1991) *For the Love of Enzymes: The Odyssey of a Biochemist.* Harvard University Press, p. 207.
17. Lehman IR, Richardson CC. (1964) The Deoxyribonucleases of *Escherichia coli*. IV. An Exonuclease Activity Present in Purified Preparations of Deoxyribonucleic Acid Polymerase. *J Biol Chem* **239**: 233–241.
18. Brutlag D, Kornberg A. (1972) Enzymatic Synthesis of Deoxyribonucleic Acid. XXXVI. A Proofreading Function For the 3′→5′ Exonuclease Activity in Deoxyribonucleic Acid Polymerase. *J Biol Chem* **247**: 241–248.
19. Friedberg EC, Walker GC, Siede W, *et al.* (2006) *DNA Repair and Mutagenesis*, 2nd edition, ASM Press, Washington, DC.
20. Ibid.
21. Kelly RB, Atkinson MR, Huberman JA, Kornberg A. (1969) Excision of Thymine Dimers and Other Mismatched Sequences by DNA Polymerase of *Escherichia coli. Nature* **224**: 495–501.
22. Letter from Arthur Kornberg to Francis Crick, July 3, 1969. Reproduced with permission from the Department of Special Collections, Stanford University Libraries.
23. John Cairns — Biography. http://library.cshl.edu/oralhistory/speaker/john-cairns/

24. Horwitz R. (2008) Tribute to Arthur Kornberg, Stanford University School of Medicine, January 25, 2008, p. 36 (private publication).
25. What Makes DNA Duplicate? NEWS AND VIEWS, *Nature* **224**: 1151–1152 (1969).
26. Polymerases. Multifarious Molecules. *Nature* **225**: 586–587 (1970).
27. AK Interview with Sally Smith Hughes, Program in the History of the Biosciences and Biotechnology, Biochemistry at Stanford, Biotechnology at DNAX, Arthur Kornberg, 1997, p. 64.
28. Kornberg A. (1991) *For the Love of Enzymes: The Odyssey of a Biochemist*. Harvard University Press, p. 217.
29. Smith DW, Schaller HE, Bonhoeffer FJ. (1970) DNA Synthesis *in vitro*. *Nature* **226**: 711–713.
30. Knippers R, Stratling W. (1970) The DNA Replicating Capacity of Isolated *E. coli* Cell Wall-Membrane Complexes. *Nature* **226**: 713–717.
31. Knippers R. (1970) DNA Polymerase II. *Nature* **228**: 1050–1053.
32. Kornberg A. (1991) *For the Love of Enzymes: The Odyssey of a Biochemist*. Harvard University Press, p. 218.

CHAPTER TEN

Like Father Like Sons

As mentioned earlier, Arthur and Sylvy Kornberg raised three sons, Roger David, Thomas (Tom) Bill, and Kenneth (Ken) Andrew, born in 1947, 1948 and 1950, respectively — years when Arthur was at the NIH in Bethesda, Maryland. At the time of this writing all three enjoy successful professional careers, Roger and Tom as academic scientists at Stanford University and the University of California at San Francisco respectively, and Ken as an architect with special expertise in designing science laboratories.

Though Kornberg devoted scant attention to his personal life in his autobiography *For the Love of Enzymes: The Odyssey of a Biochemist*, he notably included a brief section entitled *Biochemistry, a Family Affair*, much of which is devoted to accounts of his children and their exposure to science through him and his wife Sylvy. A heavy diet of science from their parents starting at very young ages was of course predictable — but never resented. Discussions about biochemistry and molecular biology prompted by Arthur and Sylvy were standard fare at the dinner table, and all three boys were frequent visitors to their father's research laboratories at Washington University and Stanford, both after school, on weekends and in later years during summer breaks. Science was in their blood at early ages — and they genuinely loved it.

"Our children were often with us in the lab after school let out and on weekends," Kornberg wrote. "It was a busy, congenial atmosphere in which postdoctoral fellows and others gave them fond attention. When asked at age nine what he wanted for Xmas Roger unhesitatingly answered: 'A week in the lab!'"[1] From their earliest ages, even when still

Arthur Kornberg displaying his fondness for children.

in diapers, Arthur took one of his sons with him on science-related trips. Later on there were more extensive voyages, even weeklong trips when he was a visiting professor somewhere. "Even though I was busy and consumed by my preoccupation, my sons will tell you that I always had time to be with them," he told an interviewer.[2]

> These trips were great for them and me. With three boys so close in age and highly competitive, it was good to take one to be the center of my and other peoples' attention. Now they're very close friends, devoted to each other. There's nothing more a parent can wish for.[3]

Nor was there any apparent boredom suffered by his sons during these adventures. Kornberg made sure that the chosen offspring was suitably engaged with schoolwork when left alone — and he devoted as much of his free time as possible to be at their side. In anticipation of a visit to the Medical College of Georgia in March 1961 Kornberg wrote to his host as follows:

> I wonder whether I might impose upon you to leave a little time free so that I can take advantage of doing some sightseeing in and around

Augusta with my son. I would appreciate having a room with a TV set for Tom's amusement, although he is planning to have a good deal of school work and reading with him to occupy the hours while I am away.[4]

Later that year Kornberg contacted the organizer of a major symposium in Houston, Texas, to which he had been invited to speak, requesting permission to have young Roger accompany him.

I was wondering whether it would be feasible to bring along with me to the conference my son Roger, a high school student, Kornberg wrote. He is very interested in science and has had enough exposure to biochemistry at home and in the laboratory to enjoy hearing some of the proceedings of the meeting.

But reading between the lines one gains the distinct impression that Arthur was firmly persuaded by Roger's aptitude for biochemistry.

Roger had a princely preparation for biochemistry, spending successive summers after high school and during college at Stanford learning nucleic acid enzymology with Paul Berg, bacterial genetics with Charles Yanofsky in the Department of Biological Sciences and organic chemistry with Carl Djerassi in the Department of Chemistry. When Charles Richardson came to Stanford as a postdoctoral fellow in 1961, Roger [aged 14] taught him how to purify DNA polymerase ---- but was embarrassed at not knowing what ammonium sulfate was![5]

Roger Kornberg, by all accounts as brilliant as his father, if not more so, displayed a keen aptitude for laboratory work at an early age. A letter from Arthur to Ernie Simms, his technician in the Department of Microbiology at Washington University, notes that "this summer Rog (aged 14) spent a good deal of time in the lab with me and most of it was spent on the early steps of the polymerase purification. He enjoyed it and did well at it."

Roger attended Harvard College where, having already acquired considerable hands-on experience in the laboratory, he devoted himself to taking "the advanced courses in chemistry, physics and mathematics that his father never enjoyed — and always missed."[6] Upon returning to the California Bay Area Roger "lived at home, where lab problems and progress were common dinner topics."[7] "Both my parents had fine scientific minds and taught by example how to approach questions and problems in a logical, dispassionate way," Roger commented. "Science was a part of dinner conversation and an activity in the afternoons and on weekends. Scientific reasoning became second nature. Above all, the joy of science became evident to my brothers and me."[8]

Both Arthur and his wife Sylvy expressed complete trust in their children and general confidence in their abilities and maturity. In his autobiography Kornberg wrote:

> They shared access to our bank account from the time they entered college [Ken Kornberg relates that he received a credit card at the age of 13 while in high school] and reciprocated our feelings toward them. Perhaps this shared sense of trust helped them cope with the turbulence of the '60s and '70s without turning to drugs or dropping out to "search for identity." Beyond the deepening friendship that came with maturity, what has given me even more comfort is their intense loyalty to one another.[9]

As a graduate student in Harden McConnell's laboratory in the Stanford Department of Chemistry Roger applied himself to the study of biological membranes using nuclear and paramagnetic resonance techniques, work that led to the discoveries of phospholipid flip-flop* and lateral diffusion. Keen on learning about X-ray diffraction as a methodological tool for the physico-chemical analysis of macromolecules, in 1972 Roger progressed to a three-year postdoctoral training stint with Aaron Klug at the Medical Research Council Laboratory of Molecular

---

*Transverse diffusion or flip-flop involves the movement of a lipid or protein from one membrane surface to the other. Unlike lateral diffusion, transverse diffusion is a fairly slow process due to the fact that a relatively significant amount of energy is required for flip-flopping to occur. [http://en.wikibooks.org/wiki/Structural_Biochemistry/Lipids/Membrane_Fluidity]

Biology (LMB) in Cambridge England. Klug (who would win a Nobel Prize in Chemistry in 1982) was a leading crystallographer and was also responsible for the application of Fourier methods to electron microscopy and image processing.

While in search of a project that would lend itself to X-ray diffraction analysis Roger met Mark Bretscher, apparently then the only scientist at the LMB interested in membrane structure. But rather than discuss membranes Bretscher suggested that Kornberg read a recent paper by Francis Crick entitled "A General Model for Higher Organism Chromosomes." The paper included a diagram showing a loop of DNA crossed by a dashed line, to symbolize a histone molecule. Klug in turn produced a sheaf of papers on the X-ray analysis of chromatin, a nuclear constituent known to contain about equal amounts of histones and DNA — and encouraged Roger to pursue the problem. "He warned me, however, that it was a 'messy' one," Roger commented.[10] "'Notorious' might have been a better word," Roger noted in a biographical statement surrounding his 2006 Nobel award.

> Many had succumbed to the allure of the problem, with its potential for insight into genetic chemistry, only to be frustrated by the intractability of the histones. These proteins were, on the one hand, surprisingly simple — and on the other hand, hopelessly complicated.[11]

Regardless, Roger made impressive strides in characterizing chromatin and its organization into discrete structural units called nucleosomes.

Nucleosomes comprise the fundamental repeating units of eukaryotic chromatin, which is used to pack the large eukaryotic genomes into the nucleus while still ensuring appropriate access to it. In mammalian cells ~2 meters of linear DNA are packed into a nucleus of roughly 10 μm diameter. Nucleosomes are folded through successively higher order structures to eventually constitute chromosomes. This both compacts DNA and creates an added layer of regulatory control, which ensures correct gene expression. Nucleosomes, first observed as particles in the electron microscope by Don and Ada Olins in 1974, are also thought to carry epigenetically inherited information in the form of covalent

Roger Kornberg.

modifications of their core histones. Their existence and structure as histone octamers surrounded by approximately 200 base pairs of DNA was first proposed by Roger Kornberg.

Upon returning to the US in 1976 Roger assumed a junior faculty position in Harvard's Department of Biological Chemistry and two years later returned to Stanford as a faculty member in the Department of Structural Biology, where he presently pursues his scientific interests on the regulation of transcription in living cells with emphasis on the reconstitution of the process with more than 50 purified proteins and structural studies of the 50 protein complex, as well as studies on chromatin remodeling required for transcription of the DNA template in living cells. At Stanford he mentored Yahli Lorch, a graduate student from Israel who subsequently became his wife and who also occupies a faculty position in the Department of Structural Biology. The couple have three children.

In the mid-1990s Roger Kornberg was unraveling the molecular components of a transcription initiation factor called TFIIH (transcription factor IIH). In collaborative studies with the author it was established that the subunits of TFIIH are also essential for nucleotide excision repair of damaged DNA in eukaryotes, an observation suggestive of a regulatory link between transcription and DNA repair.

As mentioned earlier, in 2006, a year before his father's death, Roger was awarded the Nobel Prize in Chemistry "for his studies of the molecular basis of eukaryotic transcription," making Arthur and him the sixth father/son pair to win Nobels.* On the day the prize was announced Roger was scheduled to fly to Pittsburgh to receive the Dickson Prize in Medicine. When he called the airline to cancel his flight the travel agent enquired why he was doing so. "Well, I just won the Nobel Prize in Chemistry," he sardonically replied, as if this was a well-worn explanation for canceling flights![12]

****

Music became a supplement in the lives of Arthur Kornberg's sons when at the insistence of their mother Sylvy all three were enrolled in lessons. Roger learned the violin and Ken the viola, though neither exhibited a serious interest in music as career choices. Tom on the other hand, who learned to play the cello, became a fervent and dedicated musician, to the extent that by the time he reached high school his passion for music in general and for the cello in particular precluded any special attention to science. As presently recounted, had Tom not suffered a physical disability that curtailed his capacity for playing the cello he might well have pursued music as a career.

At the time that Tom began seriously cultivating his interest in the cello, the classical music scene in the San Francisco Bay Area was poorly developed and finding experienced and adept teachers was a struggle.[13] The person most highly recommended as a mentor by those in the know

---

*The other five father/son teams were: Aage N. Bohr's 1975 prize followed father Niels' in 1922; Hans von Euler-Chelpin (1929) and Ulf von Euler (1970), Manne Siegbahn (1924) and Kai M. Siegbahn (1981), J.J. Thomson (1906) and George Paget Thomson (1937), William Henry Bragg and his son William Lawrence Bragg (1915). Notable too is the Nobel's lone father and daughter duo: Pierre Curie, who shared the 1903 prize in Physics, and Irène Joliot-Curie, who shared the 1935 prize in Chemistry. [http://news.stanford.edu/news/2006/october11/kornberg-101106.html]

was the famous American cellist Leonard Rose at the Julliard School in New York, who at age 21 was principal cellist of the Cleveland Orchestra and at age 26 occupied that position with the New York Philharmonic Orchestra. Rose (who died in 1984) was the pre-eminent concertizing American cellist of his generation. He played concertos with all the major orchestras in the country, gave recitals in all the major concert halls, and he recorded with the finest conductors and orchestras of his day. He was also widely recognized as the pre-eminent teacher of his generation and his students populated the cello sections of many if not all the major orchestras.[14]

Through the auspices of the principal cellist of the San Francisco Symphony Orchestra Tom Kornberg, then about 15 years old, was able to contact Rose and arrange to meet and audition for him. "That year the geographically closest that Rose was going to be to the Bay Area was Tucson, Arizona," Tom related.

> So I went to Tucson and auditioned for him in a hotel room. I don't think it was mere coincidence, but my father gave a seminar at the University of Arizona at the same time! I arrived at the hotel first and when he registered he did so as Tom Kornberg's father![15]

Tom Kornberg.

Tom's audition with Rose was sufficiently impressive to the impresario that he expressed his willingness to accept Tom as a pupil at the famed Julliard School of Music, an undeniable tribute to Tom's musical potential. The offer naturally required that Tom relocate to New York after completing high school, a nuance readily accommodated by gaining admission to Columbia University.

As keen as he was to pursue a career as a cellist, Tom heeded his father's insistent advice that while studying music at the Julliard, a facility conveniently close to Columbia University, he prepare himself for an alternative career as a scientist — in case playing the cello did not work out as a career choice. "I was basically a registered full-time student at Columbia College where I took courses in biology, chemistry and physics," he stated. "But I was simultaneously taking music classes — chamber music, orchestra, music theory and cello lessons — at the Julliard."[16]

While at the Julliard Tom befriended the celebrated pianist Emanuel Ax, a contemporary about six months younger than him. "We met as beginning undergraduates when we were both enrolled at Juilliard and Columbia College," Tom related.

> We got to know each other quite well and played together often. Manny (as Ax is referred to by his friends) frequently joined me for my lessons with Leonard Rose in order to play the piano accompaniments to sonatas or concertos that I was studying, and together with the violinist Paul Biss (heralded as one of the most influential violin teachers of the 20th century) we enrolled in a chamber music performance class.[17]

"Manny accompanied many string players at their lessons, not only for Rose students but also for violinists studying with Ivan Galamian (another influential violin teacher, of Iranian-Armenian descent) and Dorothy Delay (an American violin instructor at the Juilliard)," Tom stated. "Yo-Yo Ma, a French and American cellist who was born in Paris to Chinese parents, but spent his schooling years in New York City — and was a child prodigy, performing from the age of five — was one of the other cello students with Rose at the time."[18] Over the years Emanuel Ax and Tom Kornberg have maintained contact, enabled by Ax's visits to the Bay Area for recitals and appearances with the San Francisco Symphony Orchestra. "He no longer stays at our house, but invariably visits, often to practice or play chamber music," Tom commented.[19]

The cello can take a harsh toll on performers' fingers, especially the index finger of the left hand, which is used for percussion. "If you examine the hands of any cellist you will see that they have developed large callouses," Tom stated. Performing the cello for many hours on a daily basis also takes a huge toll on the back and neck. In Tom's case these occupational hazards manifested in the eruption of painful lesions in his left index finger, diagnosed as microneuromata. "In those days musicians as well as dancers who developed these sorts of physical problems were rarely treated by the medical community," Tom said, "in part because physicians were then reluctant to deal with them, often considering them as psychological issues. Nowadays well-developed medical specialties are in place for treating such maladies."[20]

Tom's infirmity was not successfully treated until he completed his university education. In the two-year interim he was totally unable to play the cello, an experience that presumably caused him untold frustration. Eventually, Robert (Bob) Chase a hand surgeon at Stanford successfully operated on Tom. In late August 1975, Tom joyfully informed Chase:

> I am enjoying being able to play the cello again and for the most part without discomfort. -------- Though I am still experimenting to learn how much the finger can take I am optimistic that it will permit me to resume some serious playing.[21]

"It took decades for the consequences of that surgery to fully resolve," Tom related in 2014. "But for the past five years I have been able to play, for the most part without discomfort, and now play as often as I wish." Indeed, Tom Kornberg's present habit is to devote a full hour and a half each day to the cello. "I absolutely love it," he joyfully exclaimed.[22]

In early 2008 Tom Kornberg had an opportunity to display his prodigious talent as a cellist on the Stanford campus when, accompanied by three other musicians on the violin, viola and piano he offered a musical tribute in memory of his recently deceased father. In the spring of 2010 during one of Emanuel Ax's visits to the San Francisco Bay Area, Ax and Tom Kornberg rendered a series of duets at Stanford, an event sponsored by the Department of Biochemistry. "What a thrill for me to play with Emanuel," Tom exclaimed. "The only other cellist he had played these particular pieces with was Yo-Yo Ma!"[23]

When it became apparent to Tom that he would have to forgo a musical career as a cellist he began exploiting Arthur's providential advice to prepare for an alternative career. Having wisely registered for appropriate courses at Columbia he, like his father and his older brother Roger, turned to biochemistry and molecular biology. After obtaining his PhD at Columbia in 1973 Tom pursued postdoctoral training at Princeton University and subsequently at the Cambridge MRC Laboratory of Molecular Biology — following immediately on the heels of his brother Roger, who had just completed a stint there.

Tom was well aware of the work ongoing in his father's laboratory and of the controversy surrounding the implications of the viable *E. coli* mutant isolated by John Cairns and his colleague Paula DeLucia. At that time he was providentially taking an undergraduate biochemistry course offered by Malcolm Gefter, a Columbia University faculty member.

"We were on a one-day ski trip near New York city one time and Malcolm casually suggested that it might be an interesting — and certainly an amusing experience, if I joined his laboratory to work on the Cairns mutant with him, which was not something he was actively engaged with then," Tom related.

> At first we jokingly laughed it off. But a few months later, when I had to finally stop playing the cello I decided to take Malcolm up on that offer. Many of the science courses I was taking at Columbia had laboratory exercises associated with them. But every time I signed up for one I simply informed the professor that I was a Julliard student and could not devote the time required for the lab courses. So I graduated from Columbia without taking a single lab class — and knew essentially nothing about working in a biochemically oriented research laboratory.[24]

As Arthur Kornberg's devoted and loyal son, Tom was sensitive to the pejorative editorials that had been written in *Nature* in 1969 and 1970 about the DNA polymerase discovered by his father suggesting that it was a mere "red herring" with regard to DNA replication. "At that time my father was on sabbatical in Cambridge, England working on membranes and on sporulation," Tom recounted.

> There was no e-mail of course, and it wasn't then financially trivial to simply pick up the phone to talk to him any time I liked. But we did speak occasionally and he cautioned me about embarking on work with the Cairns mutant. He told me that several major DNA replication figures had examined the mutant and hadn't found anything interesting. He did what he could to discourage me. But I was the classical case of someone who didn't know enough to listen![25]

Tom petitioned Gefter to provide him with a small area of bench space. "I learned how to make cell extracts and carry out DNA polymerase assays," he related.

> I was hoping and actually assuming that I would be able to show that DNA polymerase I was in fact present in the Cairns mutant, but in a form that made it difficult to detect. So I decided to go on a mission to show that the enzyme was really there.

"Not being a DNA replication nut I was not prejudiced to do things the way they were typically done," Tom continued.

> I miniaturized the assay, making incubation times much shorter and using a DNA substrate with a much higher specific activity than anyone had done before. I was able to improve the conditions for detecting DNA polymerase and was eventually able to detect a smidgen of activity. One wouldn't even pay any attention to this if one were using a wild type strain. I was convinced that I had rediscovered DNA polymerase I, but eventually I found that some of the properties of this activity were different from that of Pol I, indicating the presence of a novel DNA polymerase. On the one hand I was disappointed. But on the other hand it was nice to keep it in the family![26]

A July 1970 letter to Arthur Kornberg (who was then on the tail end of his sabbatical in England) from acting chairman of the Stanford

Department of Biochemistry Paul Berg shared the essence of conversations that he had with Tom Kornberg about the work he had been engaged in and revealed some of the "secrets" about his success with observing DNA polymerase activity. "I think the crucial point of what he has found is that starting with the Cairns mutant, he can isolate a DNA-dependent DNA polymerase," Berg wrote. "It requires all four triphosphates, but is not neutralized or affected in any way by the anti-Kornberg polymerase."

> "It seems to me quite possible that he has learned how to extract the classical polymerase from the Cairns mutant in such a way as to preserve its activity," Berg wrote. "He told me that when he uses standard extraction procedures on the Cairns mutant he finds no polymerase activity in the extract. His activity is found only when he breaks open the cells by pressure disruption. He has never used his extraction method on the wild type to see whether he can detect both the classical polymerase and the activity he has studied. --
> -- If Tom publishes this, that will surely give the editors of *Nature* a headache. They won't know which Kornberg is doing what!"[27]

Tom Kornberg and Malcolm Gefter published their startling results in *Biochem. Biophys. Res. Comm.* in late 1970. Considering all likely possibilities for their results they thoughtfully wrote: "the activity that we have described can be the result of: (1) a mutant DNA polymerase, (2) a mutant DNA polymerase complexed with another protein or membrane component, or (3) a new DNA-synthesizing enzyme." As mentioned in the previous chapter, in December 1970 Rolf Knippers reported the detection of a novel membrane-associated DNA polymerase activity in the Cairns mutant that he called DNA polymerase II. At about the same time, similar results emerged from Friedrich Bonhoeffer's laboratory. These publications eclipsed that by Tom Kornberg and Malcolm Gefter by a matter of mere months, and for all practical purposes the independent contributions to the literature from the laboratories of Knippers, Bonhoeffer and Gefter must be considered contemporaneous. In April 1971, a few months after their report in *Biochem. Biophys. Res. Comm.*, Tom and Malcolm Gefter published a paper in the *PNAS* in which they announced the purification and properties of DNA polymerase II of *E. coli*. In the discussion section they noted the existence of a third DNA polymerase. "During the course of purification, a second polymerase activity present in polA [mutant] extracts has been observed," they wrote. --------- "It can be clearly distinguished from polymerases I and II. ------ We must however, be cautious in assuming that [this] is a distinct new enzyme."[28]

In December 1971 the Gefter research group published another paper in the *PNAS*, this a multi-authored paper that included Tom Kornberg, which unambiguously identified DNA polymerase III of *E. coli* and demonstrated that the enzyme is the product of an essential gene called *dnaE*.[29] "From these results we conclude that DNA polymerases II and III are independent enzymes and that DNA polymerase III is an enzyme required for DNA replication in *Escherichia coli*," the authors stated in the abstract.[30] As discussed in a later chapter, subsequent work in Arthur Kornberg's laboratory unequivocally defined *E. coli* DNA polymerase III as the true replicative polymerase.

In his lengthy 1997 interview with Sally Smith Hughes, Arthur Kornberg reminisced about his son Tom's courage and perseverance in undertaking this defense of his father's opinions about DNA polymerase I.

> Here is your child who loved the cello, who was devoted to it without any parental pressure, who would spend many hours in the day, four, five, six hours practicing a few notes to get them right. Tom was very driven and motivated, and gifted enough so that he was accepted by the premier cellist in the country, Leonard Rose, as a student, and at Juilliard. He was doing very well and had the mental stamina to cope with Yo-Yo Ma, who was his classmate. Then he developed a lesion in his left index finger, which was like a major athlete losing his arm or leg and being unable to function. It was very painful, and he simply couldn't use his finger. It looked like the termination of his musical career at which he had labored more than ten years. And to this day, he is a very fine cellist and loves music. That's one side of it. The other is that unlike Roger, his older brother, or Ken, his younger brother, Tom was so devoted to music that he hadn't had any lab experience. In the place of music, he was going to try to do what Bob Lehman, Charles Richardson, and others who were very practiced at research on DNA replication and

[DNA] polymerase were unable to do. I thought he'd waste his time, get frustrated further. So my advice was, "Tom, these people have tried to find the other enzyme in the Cairns mutant; they haven't been able to do it." So yes, I discouraged him.

Then he went to work in a lab at Columbia where he was also a student, and where Malcolm Gefter gave him lab space — and I don't know the details — modest encouragement. Within weeks, he had manipulated cells and extracts, and demonstrated that, yes, in these mutants there is another activity. Then months later, he found that he could detect still a third activity which proved to be the ultimate polymerase. I know that Jerry Hurwitz, who can be a very accomplished and decent person but can also be carping and critical, ridiculed the data on which Tom was basing the claim that there was a third polymerase. Tom turned out to be right. What were my feelings? Certainly pride. And why did Tom do it? I think he did it out of loyalty. He wanted to show that his father, unlike what was being said in lectures and hallways, was not misleading the world as *Nature New Biology* was claiming, but that there was a likelihood that for some reason in this mutant my DNA polymerase was being masked or somehow inhibited.[31]

"Though long since forgotten, my father was affected by the implication of the *Nature* editorials," Roger Kornberg related.

He was also very touched by Tom's concern when he came under assault by *Nature*'s implication that the Nobel Prize for DNA polymerase was perhaps undeserved. Francis Crick put it very well. When my father was there on sabbatical in 1969 and asked Francis what he thought about all the ballyhoo Francis's response was that regardless of whether or not DNA polymerase I was involved in DNA replication, the one(s) that is(are) will turn out to be very similar biochemically.[32]

On July 5, 1970 Arthur Kornberg wrote to his son Tom. "To begin with you must know that I am proud of what you are doing," he stated. "Revelations about DNA polymerase will continue for many years and I'd rather you made them than anyone else, including myself."[33] But not withstanding the sentimentality behind his son's heroic efforts Kornberg could not resist offering a friendly denunciation.

> Since you seem inclined to think that an enzyme different from DNA polymerase is responsible for your synthetic activity, let me set down the arguments against it. ------------------ It seems to me that a reasonably strong case can be made that your activity is DNA polymerase complexed with another component (a subunit of the native 'replicase'.)[34]

These sentiments were followed by a healthy dose of fatherly advice.

> I think it would help you, and certainly me, if you took an hour or two to set down on paper a brief summary of what you've done — your interpretation and plans. This would give you a refreshing pause and me a better chance to respond usefully. As someone who hates to write reports, I warn you that failure to stop frequently to prepare such summaries is one of the most wasteful omissions in research. I never have found that I wasted time in such an exercise. Putting things down on paper forces you to explain and to interpret things in a way that can easily be ignored in thinking and in conversation.[35]

"I arrived at the Laboratory for Molecular Biology (LMB) in 1972 and the spring and summer of that year I attended a summer school in Spetsai, Greece" Roger Kornberg related.

> John Cairns was there. Within a few minutes of arriving I spoke to him and defended my father's work. And someone — I can't remember who, came up to me and said: "You don't have to defend Arthur Kornberg's work," implying I assume, both that his work did not require defense and that if it did he was eminently capable of doing so himself. I didn't realize I was doing so. But it was instinctive. When Tom went into Gefter's lab he had a similar notion. So we were both sensitive to that nuance.[36]

Perhaps forgetting or never thinking that the transcript of his interview with Sally Hughes was a public document, Arthur Kornberg could not resist taking a shot across Gefter's bows — that he might have done without. "Gefter received a prize for that discovery, and didn't deserve it," Kornberg stated. "It was Tom's work which Gefter then expanded; he was instrumental in the genetics that established that this

Ken Kornberg.

new polymerase was a truly novel protein encoded by a different gene."[37]

Tom Kornberg resides in San Francisco with his wife Jody, a practicing psychotherapist, and his two sons Ross and Zac.

****

The youngest of Arthur Kornberg's sons, Ken, was born in Bethesda, MD, at a time when his parents were scientists at the NIH. His exposure to nucleic acids was amusingly evident at an early age. When his two brothers once enquired what acid to use to remove cement spots from their house porch, Ken gallantly suggested the only one he knew: "nucleic acid!" Influenced by the scientific atmosphere that permeated the Kornberg household, Ken spent a summer in his father's laboratory while at high school, and after completing his high school education he pursued a research stint in the *Instituto Di Genetica* at the University of Pavia in Italy.[38] In 1967, at the age of 17, Ken also spent several months in a marine biology lab at the Scripps Institution of Oceanography. He was equipped with a surfer's "woody" station wagon and scoured the beaches for new species of diatom. "That summer I discovered a new diatom, developed a lifelong love of the ocean — and met Junior Miss La Jolla! I thought science was great," he enthusiastically commented.[39]

When he enrolled at Stanford in 1968 Ken vacillated between music (apparently being more than proficient on the viola) and marine biology as his major, but it wasn't long before he was lured away by the beautiful drawings and models he observed when taking courses in the Stanford Department of Architecture.* Architecture, particularly that of science buildings and laboratories, became his chosen vocation. Ken began his career in several Bay Area architectural firms designing houses, department stores and junior colleges. He had the added bonus of meeting his wife Veronica at an architectural firm in San Francisco. He opened his first independent architectural office in Del Mar, California, where his first project was to remodel a laboratory in the Stanford genetics department. This undertaking was completed in 1980, a time when the biotechnology industry was beginning to take off in San Diego and the Bay Area. The Stanford engagement led to several architectural assignments in La Jolla, at Cal Tech, Stanford, and The University of California at San Francisco. In 1986 Ken opened a new office in Menlo Park. Since then Kornberg Associates has completed 400 biomedical research projects on four continents. Ken and his wife Veronica have three children, Jessica, Sophie and Zoe. Accordingly, prior to his demise in 2007 Arthur Kornberg enjoyed the bliss of eight grandchildren from his three sons.

References

1. Kornberg A. (1991) *For the Love of Enzymes: The Odyssey of a Biochemist.* Harvard University Press, p. 173.
2. Hargittai I. (2002) Arthur Kornberg. In: *Candid Science, II. Conversations with Famous Biomedical Scientists.* Imperial College Press, p. 65.
3. Ibid.

---

*The Department of Architecture at Stanford University was terminated during Ken's junior year, but he was allowed to complete his studies and graduated with a BA degree in architecture and a MS degree in Engineering. At the time of this writing Stanford hosts an Architectural Design Program.

4. Letter from Arthur Kornberg to Knowlton Hall, March 16, 1961. Reproduced with permission from the Department of Special Collections, Stanford University Libraries.
5. Kornberg A. (1991) *For the Love of Enzymes: The Odyssey of a Biochemist*. Harvard University Press, p. 174.
6. Ibid.
7. Ibid.
8. Conger K. (2006) Roger Kornberg Wins the 2006 Nobel Prize in Chemistry. Stanford Report, Oct. 4, 2006. [news.stanford.edu/news/2006/october11/kornberg-101106.html]
9. Kornberg A. (1991) *For the Love of Enzymes: The Odyssey of a Biochemist*. Harvard University Press, p. 177.
10. The Nobel Prize in Chemistry 2006. http://www.nobelprize.org/nobel_prizes/chemistry/laureates/2006/
11. Roger D. Kornberg–Biography/ http://www.nobelprize.org/nobelprizes/chemistrylaureates/2006/kornberg-bio.html
12. http://news.stanford.edu/news/2006/october11/kornberg-101106.html
13. ECF Interview with Tom Kornberg, Jan. 2014.
14. Leonard Rose. http://www.cello.org/cnc/rose.htm
15. ECF Interview with Tom Kornberg, Jan. 2014.
16. Ibid.
17. Ibid.
18. Ibid.
19. Ibid.
20. Ibid.
21. Letter from Tom Kornberg to Robert Chase, Aug. 30 1975. Reproduced with permission from the Department of Special Collections, Stanford University Libraries.
22. ECF Interview with Tom Kornberg, Jan. 2014.
23. Ibid.
24. Ibid.
25. Ibid.
26. Ibid.
27. Letter from Paul Berg to Arthur Kornberg, July 14 1970. Reproduced with permission from the Department of Special Collections, Stanford University Libraries.

28. Kornberg T, Gefter ML. (1971) Purification and DNA Synthesis in Cell-Free Extracts: Properties of DNA polymerase II. *Proc Natl Acad Sci* **68**: 761–764.
29. Gefter M, Hirota Y, Kornberg T, *et al.* (1971) Analysis of DNA polymerase II and III in Mutants of Escherichia coli Thermosensitive for DNA Synthesis. *Proc Natl Acad Sci* **68**: 3150–3153.
30. Ibid.
31. AK Interview with Sally Smith Hughes, Program in the History of the Biosciences and Biotechnology, Biochemistry at Stanford, Biotechnology at DNAX, Arthur Kornberg, 1997, p. 65.
32. ECF interview with Roger Kornberg, May 2014.
33. Letter from Arthur Kornberg to Tom Kornberg, July 5, 1970.
34. Ibid.
35. Ibid.
36. ECF Interview with Roger Kornberg, May 2014.
37. AK Interview with Sally Smith Hughes, Program in the History of the Biosciences and Biotechnology, Biochemistry at Stanford, Biotechnology at DNAX, Arthur Kornberg, 1997, p. 65.
38. Interview with Ken Kornberg, January. 2014.
39. Ibid.

CHAPTER ELEVEN

Spores — A Brief Interlude

"Those familiar with my research career are aware of my intense concentration on a single subject — the enzymatic synthesis of DNA — and the blinders I have worn to maintain this focus," Kornberg wrote in his autobiography *For the Love of Enzymes: The Odyssey of a Biochemist*. "Nearly forgotten now by us all are the eight years, in the midst of the DNA work (1962–1970), when half of my research effort was devoted to an arcane subject, the development and germination of spores."[1]

Kornberg's familiarity with spores may have been born of the death of his mother in 1939 from an infection by the sporulating organism *Clostridium birefringens* following gall bladder surgery, a tragedy that affected him deeply. His academic interest in these biological entities necessarily arose when he assumed the chairmanship of the Department of Microbiology at Washington University and acquired the responsibility of developing an appropriate and relevant curriculum for medical students. During his early years at Stanford he became interested in spores as subjects for research. "Now I could look beyond the 'bad' spores to the vast array of innocent species whose mysterious biology and biochemistry fascinated me," he wrote.[2] In 1962 Kornberg initiated an eight-year research program centered around spore biology and biochemistry that occupied 10 postdoctoral fellows and three graduate students and yielded 26 research publications that documented a number of basic features of the biochemistry and molecular biology of spores.

Kornberg initiated his spore work by "examining how DNA is stored in the spore, what replication machinery, if any, is included, and how the DNA is used when the spore is called upon to generate a new cell."[3]

However, a succession of graduate students and postdoctoral fellows branched out in other directions that focused on (among other topics) inorganic pyrophosphatase in spores of both *B. subtilis* and *B. megaterium*, purine nucleoside phosphorylase of spores of *B. cerues*, the origin of spore core and coat proteins, and protein turnover and patterns of enzyme development during sporulation of *B. subtilis*.

James (Jim) Spudich, a long-time faculty member in the Stanford Department of Biochemistry, was a major contributor to many of these studies. Spudich was well liked by Kornberg. "I'm one of those people who is totally in love with science," Spudich stated.

> As a graduate student I couldn't find enough time to be in the lab. At least once a week my wife Ana (also a trained scientist) would sleep on the couch in Arthur's office because I wanted to collect data points through the night. I didn't do that to please Arthur. It was integral to the experiments I was doing. But this mentality probably fitted Arthur's image of the dream student. I believe that more than anything else Arthur really appreciated people in his department who had a passion for science — the way he did. And I think that very early on in my relationship with him that led him to treat me almost as a fourth son. He seemed to really take a liking to me and I felt incorporated into the family. I don't think that would have happened if I hadn't fitted the theme of having a passion for science — and working like hell![4]

Spudich was not adverse to working on spore biology rather than the central research focus in the Kornberg lab — DNA replication. He viewed the fact that certain bacteria generate spores as a fascinating issue that he wanted to understand. "How does a bacterial cell generate a spore; something that is quite distinct?" he wondered.

> There was still a lot to do on DNA polymerase, but Arthur was then very keen to move on to something new and different. Arturo Falaschi (an Italian scientist) was in the lab at that time and he was beginning to look at the spore problem — so I thought that this would be a fun adventure.[5]

James (Jim) Spudich.

"The spore project was initiated by Arturo and me," Spudich confided.

> Eventually there were about six of us working on it. And we published. We didn't rock the world with those papers. But all in all it was a wonderful training experience in biochemistry, which was my intention in coming to Stanford as a graduate student in the first place.[6]

Kornberg evidently thought well enough of Spudich that he suggested that this favored graduate student remain in the department, hopefully to eventually become a junior faculty member. "I wasn't really interested in doing that because I believed it would have been hard to separate myself from this strong father figure," Spudich related.

> I felt that I needed to separate myself from him and establish myself elsewhere before considering returning to that department. But more importantly perhaps, I felt that one of Arthur's strengths — and weaknesses — was his total focus on grinding up cells, and purifying enzymes and I wanted to carve out a different niche for myself.[7]

Spudich commented on Kornberg's intense focus on enzymes, sometimes to the exclusion of the promise of other aspects of biochemistry and molecular biology. "One day I wanted to see what these spores actually looked like," he related.

We didn't have a light microscope in the department so I went to the Department of Genetics to see if I could locate one, and ran into Josh Lederberg. Josh was pleased that I wanted to actually look at something under the microscope so he lent me his very own. I took it up to the lab and while I was examining the spores Arthur walked in and asked me what I was doing. I told him that Josh Lederberg had loaned me his personal microscope so that I could see what the spores looked like. His immediate response was: "take that thing back right now!" So there was this side of Arthur that was a little myopic when it came to different approaches to things. Having attended a physiology course at Woods Hole before starting graduate school, a course that taught one about all the different approaches one could use in biological research, when I came to Stanford as a graduate student I found everything very focused. I realized that being a graduate student in the Department of Biochemistry was going to be a great learning experience. But I also realized the importance of mixing genetics and structural biology with biochemistry.[8]

Kornberg occasionally expressed relative disdain for molecular biology compared to his passion for enzymology. But in his defense it bears repeating that notwithstanding his passion "for grinding up cells and purifying enzymes," in Jim Spudich's words, he was by no means blind to the possibilities of applying enzymology to understanding more complex biological events. In the conclusion of an essay entitled "For The Love of Enzymes" contributed to the commemorative volume published in 1976 in honor of Severo Ochoa mentioned in an earlier chapter, Kornberg wrote:

If the guiding theme of earlier enzymologists was: "Don't waste clean thoughts on dirty enzymes," the motto of the future should be: "don't entertain dirty (pessimistic) thoughts about clean enzymes." We must assume that complex metabolic pathways will be reconstituted from purified enzymes just as the simpler ones have been in the past. With an understanding of how to fashion membranous vesicles and to orient their surfaces correctly, it should become possible to reconstitute organelles of considerable complexity. As a long range proposal I suggest the reconstitution of a successful viral infection — the production, in a few minutes, of several hundred virus particles from a single cell — in a molecularly defined and dispersed (cell-free) system.[9]

Regardless of the reasonably productive efforts on spore biology and biochemistry that emerged from his laboratory, Kornberg's interest in spores was relatively short lived, and in due course he abandoned this area of research.

> I came to realize how much more complex sporulation and germination were than I initially assumed. Several hundred genes are devoted to these processes, and we knew very few of them and their protein products. Only the scantiest information was available about the biochemistry and physiology of spores, and hardly any one else in the world seemed to care. The little research on spores, then and even now, was largely of a practical nature — how to destroy spores in food canning or how to use them as pesticides on crops.[10]

Then too, Kornberg became increasingly disturbed by the distraction those engaged with spores suffered from people in his research group who were continuing to work gainfully on DNA replication. Indeed, he offered little resistance to those who wanted to switch from spores to DNA replication. "Finally, after this eight-year siege, I gave up too," he stated.[11] In the early 1950s viruses, most notably those that infect bacteria, the bacteriophages, also entered Kornberg's sphere of interest. As described in the next chapter the use of phage-infected *E. coli* cells proved to be crucial in clarifying many features of the complexity of DNA replication.

References

1. Kornberg A. (1991) *For the Love of Enzymes: The Odyssey of a Biochemist.* Harvard University Press, p. 186.
2. Ibid.
3. Ibid., p. 188.
4. ECF Interview with James Spudich, Oct. 2014.
5. Ibid.

6. Ibid.
7. Ibid.
8. Ibid.
9. Kornberg A, Horecker BL, Cornudella L, Oro J. eds. (1976) For the Love of Enzymes. In: *Reflections on Biochemistry in Honour of Severo Ochoa*, Pergamon Press, p. 251.
10. Kornberg A. (1991) *For the Love of Enzymes: The Odyssey of a Biochemist.* Harvard University Press, p. 188.
11. Ibid., p. 189.

CHAPTER TWELVE

DNA Replication — The Holy Grail

The biochemical studies on DNA replication ongoing in Kornberg's Stanford laboratory were complemented by work in other laboratories around the world, a number of which incorporated the genetics of replication, studies that revealed the existence of multiple genes, and by inference the requirement of multiple proteins to successfully consummate the faithful copying of DNA in living organisms. Steady progress in the field notwithstanding, crucial questions remained to be answered, many of which Kornberg thoughtfully pondered. "How is a DNA chain started?" he asked.

> What are the replicative functions of the three distinctive DNA polymerases in *E. coli*? What are the specific functions of the numerous *E. coli* genes identified as essential for replication? What is the switch that controls initiation of a cycle of chromosome replication?[1]

Cartoons representing the linear Y-shaped structure of a replicating DNA molecule — discreetly covered by a fig leaf — became a popular way of lampooning how much was still to be learned about this vital function. In his autobiography Kornberg wrote:

> The psychological goad of this cartoon, added to my frustration with our lack of progress may have spurred me to finally realize a basic flaw in our work. It dawned on me that we would never answer the extant

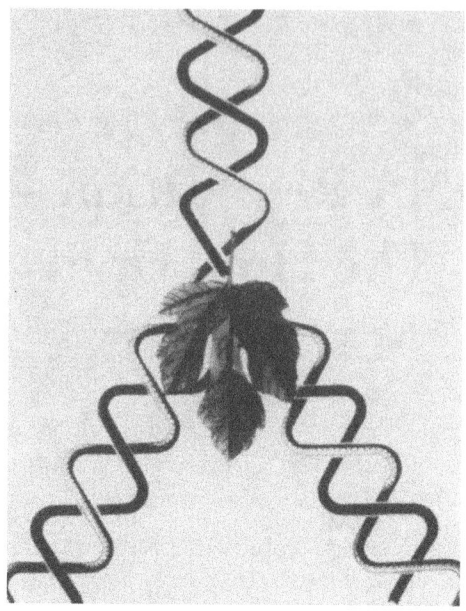

The metaphorical fig leaf hiding events at the DNA replication fork.

questions about replication with the DNA we were using as template and primer; DNA extracted from bacterial and animal cells was not an adequate substrate for the enzymes of replication.[2]

When in 1971 I finally recognized the futility of searching for replication enzymes with bacterial DNA, let alone animal DNA, I also realized that for five years we had been ignoring a proper DNA substrate, the chromosome of the tiny bacteriophages. --------------- I recalled belatedly the virtues of the intact, clean phage chromosome that four years earlier had served us in demonstrating the synthesis of infectious DNA by DNA polymerase I.[3]

Having acquired experience with the replication of a bacteriophage called M13, the single-stranded DNA circle in that bacterial virus became Kornberg's focus in his attempts to dispense with the metaphorical fig leaf. In the introduction section of a key 1971 paper published in the *PNAS* Kornberg, together with graduate students Doug Brutlag and Randy Schekman, floated several hypothetical operations that might explain how M13 DNA replication is initiated. Their hypotheses included the following:

> Since RNA polymerase starts new RNA chains, and DNA polymerase is known to covalently extend a ribonucleotide terminus during DNA synthesis, a brief transcriptional operation by RNA polymerase might provide an RNA primer for DNA synthesis. This priming piece of RNA could later be recognized and excised by nuclease action. Thus, a *de novo* initiation event catalyzed by RNA polymerase would be an essential step for the start of DNA synthesis.[4]

This is precisely what Kornberg and his colleagues were able to demonstrate by showing that rifampicin, an antibiotic known to inhibit RNA synthesis, blocked both the conversion of single-stranded DNA M13 DNA to the double-stranded replicative form — and replication of this form of DNA. They concluded their 1971 *PNAS* paper mentioned above, which bore the title "A Possible Role for RNA Polymerase in the Initiation of M13 DNA Synthesis," with the prophetic statement: "— an attractive hypothesis for the function of such an RNA chain is that which originally prompted this investigation: to act as a primer terminus for covalent extension by a DNA polymerase."[5] Not too much later the Kornberg laboratory directly demonstrated that RNA synthesis indeed initiates the *in vitro* conversion of M13 single-stranded circular DNA to its double-stranded replicative form.

The involvement of RNA polymerase in DNA replication was a surprise to many — but not to all. William (Bill) Dove at the McArdle Laboratory of the University of Wisconsin, a colleague and friend of Kornberg's who had obtained an advance copy of the manuscript published in the *PNAS* by Brutlag, Schekman and Kornberg, wrote Kornberg a letter stating:

> I find this paper very exciting because it defines the involvement of RNA synthesis in DNA replication (initiation?) in a very clean way; I expect that the study of the molecular basis of this coupling will proceed well in this case. Also, I find the explicit statement of the possibility of an RNA primer to be very constructive (i.e. testable). ------------ the average reader

may not be aware that the involvement of RNA synthesis in the initiation of DNA replication has been studied extensively in phage λ (mainly by us) and has been suggested for *E. coli* by [Gordon] Lark (1969 review). In the case of λ the evidence extends to the point implicating initiation rather than movement of replication forks. Further, it is found that the transcription necessary for initiation must lie near the origin of replication. --------- I enclose our published work on the λ case so that you can evaluate this point. ---------- I find your manuscript to be a valuable breakthrough in these studies, and in raising this point I do not intend to detract from its value.[6]

Kornberg's response was that while he was aware of this work

it seemed inappropriate to generalize and expand when we had not yet established the true function of RNA polymerase in M13 replication. In fact, we still don't know for certain that there is a priming fragment in the initiating event, and if so, how long it persists. I think we are getting closer to knowing because we've purified the extracts significantly. ----------I was stimulated to read and understand your work better. From what you have done and the indications from Lark and others, it does seem rather likely that several DNA initiations may share an RNA involvement.[7]

In a subsequent follow up of this correspondence Dove stated:

I am eager to know the outcome of your text of the RNA primer hypothesis, especially since we are not favorable towards models with such tight coupling! ---------------- I hasten to say that I am sure that your recognition of RNA polymerase involvement is completely original and independent.[8]

In 1976 Kornberg wrote a lengthy review on RNA priming of DNA replication. In concluding the article he addressed the rhetorical question: "Assuming a generality for RNA priming of cellular DNA replication, what advantage can be seen for it?" "It may be to ensure the ultra high fidelity of the indelible DNA record," was his own retort.

The proof-reading and error-correcting devices built into DNA replicative mechanisms may not function as well in removing base-pairing errors at or near the start of a DNA chain as during its growth. An RNA start, by contrast, is readily recognized as a foreign hybrid, can be specifically erased by several alternative mechanisms, and is replaced by high fidelity DNA chain growth.[9]

Another conundrum lay buried — the observation that DNA replication proceeds in a 5′→3 direction; yet the two DNA chains in a duplex DNA molecule run in opposite directions. How could these observations be reconciled? Kornberg proposed a scheme in which synthesis of the two chains in DNA transpire differently; continuously on one strand, which became known as "the leading strand," and discontinuously on the other, with chains starting repeatedly at the beginning of each discontinuous fragment, known as "the lagging strand." "In this way, the growth of the lagging strand, although opposite in direction to that of the main replication fork movement at the molecular level, seems at a gross level to be in the same direction," Kornberg surmised.[10]

Experimental support of this model was provided by the Japanese investigator Reiji Okazaki and his wife Tuneko, who demonstrated that a significant fraction of newly-synthesized DNA is indeed in short pieces approximately 1,000 nucleotides long. Okazaki's name became secured in the annals of DNA replication biology when these short segments of newly synthesized DNA became referred to as "Okazaki fragments."

Prior to this discovery Okazaki and his wife had worked as postdoctoral fellows in Kornberg's laboratory. Arthur thought highly of both scientists. Noting that "Reiji's research style is not readily gleaned from his publications," Kornberg related an amusing anecdote concerning his postdoctoral fellow's activities in the laboratory, which he jokingly referred to as the "Okazaki maneuver." "In purifying an enzyme he used a heating step; the enzyme was held in a test tube at 70° for 5 minutes," Kornberg related. "When he decided to prepare a large amount of enzyme going from a scale of 10 milliliters to several liters, how did he carry out the heating step? He simply repeated the same heating procedure several hundred times."[11] While admitting to some embarrassment at having to report this somewhat unorthodox procedure in a publication, Kornberg realized that it was nonetheless effective. "He was able to complete this step in a few hours and saw no point in wasting precious time and material in learning how to do the heating in a large beaker or flask."[12] Okazaki's laborious procedure subsequently proved to have considerable merit for others.

> Later, when another of my students purified an enzyme with a heating step and had to scale up his procedure 2,000-fold --------- I advised the Okazaki maneuver. ----------- Another student who tried heating a large volume of this enzyme in a single step lost it in a thick coagulum.[13]

Reiji Okazaki succumbed to leukemia and died in 1975. Some suggested that his leukemia was the result of exposure to radiation from the atomic bombing of Hiroshima during World War II. But a comment in several newspapers that he was a high school student in a suburb of Hiroshima when the bomb exploded and was exposed to radiation while searching for his father close to the point where it exploded is apparently inaccurate. In corresponding with a Japanese scientist in August 1975 Kornberg assured the individual that the Atomic Bomb Casualty Commission in Japan concluded that

> there had been no significant fallout and so persons entering the city a day later, as Reiji told me he did, received no radiation. Also the data of the last few years show no difference in the incidence of leukemia between bomb survivors and the general population.[14]

An obituary to Okazaki written by Kornberg in August 1975 documented another anecdote, this a tribute that Kornberg referred to as "Okazaki courage." "It is customary in my laboratory when characterizing an enzyme to set up protocols containing 10 or 12 assay tubes," Kornberg related.

> Rarely, some ambitious person will set up a 24-tube assay. Reiji set a record that will never be broken. He performed a 128-tube assay of thymidine kinase, even though each assay included a laborious electrophoretic separation of the product from the substrate. Because the pure enzyme was rather labile he felt it essential to measure all the substrate, effector, inhibitor and other parameters in the very same experiment.

The successful completion of this experiment was a feat of courage, concentration, skill and enterprise that has not been matched by anyone in my experience.[15]

****

The discovery of RNA priming during DNA replication was followed by a long, frequently tedious and frustrating period of investigation during which Kornberg and his colleagues strove to purify and characterize the multiple enzymes and other proteins required for both RNA priming and for the replication of suitably primed DNA. The many fits and starts that underscored this extended phase of Kornberg's scientific career are not described here in any detail. Readers in search of a more complete account are referred to his autobiography *For the Love of Enzymes: The Odyssey of a Biochemist.*

Kornberg devoted attention to every nuance of the formidable task of identifying and purifying the multiple proteins believed to be required for DNA replication. Months were focused on the seemingly simple task of identifying the most appropriate strain of *E. coli*, "one whose cells could be disrupted under the gentlest conditions to yield a bountiful harvest of the proteins we were seeking."[16] Early studies promised quick resolution. When the research group began purifying components required for DNA replication two primary groups of proteins surfaced: one that primed the beginning of a new DNA chain, and another that extended the chain. But as they labored to purify these fractions complexity grew as each primary fraction mushroomed further components and "the joy of uncovering the trails of so many novel proteins soon gave way to the discouragement of being unable to track down any one of them." Kornberg opined that eight different proteins might be needed to just synthesize the small amount of RNA that primes the synthesis of a DNA chain. "What were all these proteins and what was each doing?" he asked.[17]

Two major developments facilitated glimpses into the biochemical features of the initiation of replication in *E. coli*. The first involved the identification of a sequence of 245 base pairs called *oriC*, the unique origin of replication in the four million base pairs of the *E. coli* chromosome. The second was unraveling the enzyme system responsible for the replication of plasmids carrying *oriC*. "This proved far more difficult than I had imagined," Kornberg wrote. "Five people in my group spent an aggregate of 10 years without any significant progress."[18] One of them, graduate student Bob Fuller, spent two frustrating years laboring to prepare a cell extract that could replicate an *oriC* plasmid — with no discernible progress. When Kornberg began harassing Fuller to move to another project that would yield a PhD thesis Fuller stubbornly resisted. "Give me three more months," he pleaded.[19] Eventually, one day in mid-1981, while working with postdoctoral fellow Jon Kaguni, lightning struck! The pair uncovered a highly specific cell-free system for the replication of *oriC*. "It was more powerful than we ever hoped for," Kornberg exclaimed. "The gates were now opened for us to discover the parts and operations of the mechanism that initiates the duplication of a chromosome."[20]

"I never imagined that starting and completing a short circle of phage DNA could be so complicated," Kornberg wrote in his autobiography.

> In 1967, to wide acclaim, we thought we had been able to accomplish it in the test tube with just two enzymes: a polymerase to start and extend a chain and a ligase to join the ends. In 1988, after seventeen years of working on an operation the cell completes in a few seconds we are still grappling with its awesome complexity.[21]
>
> We were lucky to find DNA polymerase I. It is an extraordinary enzyme in its capacity to build a huge macromolecule under template direction and with astonishing accuracy. But then we were sentenced to wander for the next 20 years in search of the dispersed and missing pieces of the replication machine.[22]

Kornberg dubbed the complex protein assembly that primes the beginning of a new DNA chain the *primosome*. The term *DNA polymerase III holoenzyme*, a complex of more than 20 proteins, was coined to describe the "sewing machine" that assembles the DNA chains. This level of structural complexity of the DNA replication machinery was extended in Kornberg's thinking by a putative *replisome*, "a super polymerase that also includes auxiliary replication proteins, such as helicases, that unzipper the parental duplex to advance the replication fork."[23] "A more remote frontier is the final phase of replication," he wrote, "when the newly-replicated pair of chromosomes must be partitioned correctly between the two daughter cells."[24]

> Based on the insights we gained from the replication of the small phages as to how DNA chains are started, elongated, and completed, we could finally peek under the edges of the "fig leaf" covering the replication fork and begin to see some of the working parts of the replication machinery and imagine how they might drive the replication of the duplex chromosome,

Kornberg triumphantly wrote.[25] He was now able to conceptualize a discrete series of events during DNA replication. He suggested that the process begins with unwinding of the DNA duplex in advance of the replication fork by DNA helicases, thereby exposing single-stranded template DNA. Exposed single-stranded regions of DNA are coated by a single-strand binding protein, protecting the DNA from attack by ubiquitous nucleases. The multisubunit replisome containing DNA polymerase III subsequently builds the DNA chain along one template strand in a continuous fashion, constituting the so-called leading DNA strand of replication, while repeated initiations on the other strand by the primosome discontinuously generate the lagging strand. During synthesis of the lagging strand, the primosome creates a short RNA primer that is extended by DNA synthesis, producing an Okazaki fragment that fills in the available template until it reaches the previous fragment. DNA polymerase I recognizes the primer RNA on the previous Okazaki fragment as foreign, removes it and fills in the resulting gap. DNA ligase then seals the joint between the two DNA fragments.[26] A cogent function for DNA polymerase I, the enzyme that had launched Kornberg's long journey to understanding the function of the protein he had discovered in the mid-1950s, was finally revealed.

"While correct overall, the operation [of DNA replication] in molecular detail seems much too jerky and clumsy for the elegance we

208 *Emperor of Enzymes*

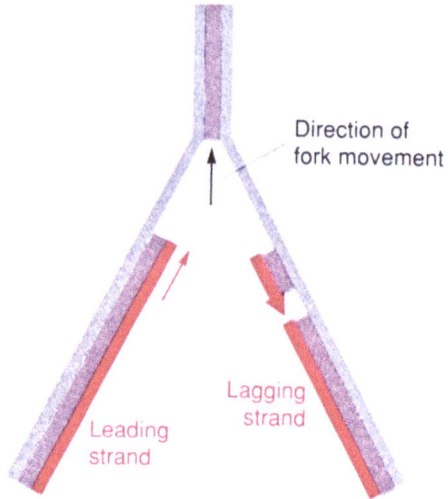

The leading and lagging strands during DNA replication.

expect of Nature's chemistry," Kornberg wrote about the sequence of events postulated above.

> We imagine a more efficient device that embraces the polymerase, the primosome and helicases — in essence a super entity, referred to as the *replisome*, that might be able to carry out the replication of both strands *concurrently* rather than spasmodically.[27]

In retrospect, Kornberg was fortunate to have selected M13 phage over ϕX174 for many of his experiments. Had he not switched to the use of M13 phage he might have missed out on the discovery of the importance of RNA polymerase in DNA replication, since replication of ϕX174 circular single-stranded DNA is not sensitive to rifampicin.

> I was prompted to do so by the unusual ways in which the cell's membranes are involved in the disassembly of M13 at the cell surface and the assembly of the phage particles --- aspects of viral infection that had intrigued me during my previous [sabbatical] stay in England, he wrote. The reason I was there is that I had felt uneasy about basic questions regarding membrane biochemistry that had arisen during the

previous six years' work on the production and germination of bacterial spores. More immediately, I wondered whether our failure to discover how DNA chains are started in replication might be due to inattention to membranes.[28]

When intent on learning more about membranes and their possible role in DNA replication Kornberg had consulted his son Roger, then a graduate student with Harden McConnell at Stanford University and well informed about membrane biochemistry. The fundamental structure of membranes was then fiercely debated between the lipid bilayer and so-called proteolipid hypotheses. During the period 1967–1972 the issue was decided in favor of the lipid bilayer, a decision significantly influenced by Roger Kornberg's magnetic resonance studies. When Arthur had asked for his son's advice as to where he might best profit from a sabbatical dedicated to learning about membranes Roger recommended working with Alec Bangham at Cambridge University, best known for his seminal research on multilamellar phospholipid vesicles, entities that came to be popularly referred to as liposomes.[29] "Bangham was a superb British eccentric, the likes of Peter Mitchell, J. D. Bernal, the Mitchisons, and so many remarkable others in British scientific circles," Roger commented. At the time of Bangham's death in March 2010 PubMed had listed more than 35,000 articles on liposomes![30]

Kornberg elected to spend the first six months of his sabbatical working alongside Roger in Harden McConnell's laboratory at Stanford, and the subsequent six months in Cambridge, where he occupied a desk in an office shared by Francis Crick and Sydney Brenner. He learned a lot about membranes, though nothing that piqued his interest with respect to their possible role in DNA replication. Kornberg wrote:

> I decided that in my future work I would continue to focus on replication and look for proteins that might operate from a membrane base rather than switch to studies of the phospholipid components of membranes. I also wanted to get back to the basics of replication enzymology.[31]

While in Israel in 2000 Kornberg delivered a lecture in Jerusalem in which he presented 10 maxims that he had gleaned from his years of dissecting the many intricacies of DNA replication. These were subsequently published in the *Journal of Bacteriology* as an invited Guest Commentary entitled "Ten Commandments: Lessons From the Enzymology of DNA Replication." Perhaps his sojourn in the Holy Land moved him to dub his article with a religious tone! Regardless, the end result is an informative analysis of the do's and don'ts in fractionating cell-free extracts, with likely application to areas of enzymology beyond DNA replication.

> In this Commentary I will give an anecdotal account of the lessons I learned from my attempts to resolve and reconstitute biological events in DNA replication and reflect on how these lessons may still apply to solving the current problems of growth and development and the aberrations of disease,

Kornberg wrote in his introduction.[32] The article offers the added bonus of explaining the rationale of many experimental nuances that are not typically addressed when documenting experimental results in the literature.

While this "holy" list of THOU SHALTS (as he referred to his catalog of commandments) related specifically to the enzymology of DNA replication, Kornberg believed that the lessons he offered had application to many if not all areas of biochemical research. In addition to edifying and defending the virtues and power of biochemistry applied to cell-free systems, a primary significance of this quaintly crafted article lies in the opportunity it afforded Kornberg to retrospectively revisit the manifold examples of his enzymological successes over the years in his quest for understanding how cells replicate their genomes through his devoted dedication to the prokaryote *E. coli* — and the viruses for which it serves as host. Kornberg's "Ten Commandments: Lessons From the Enzymology of DNA Replication" reads as follows:

1. ***Rely on Enzymology to Clarify Biological Questions*** Based on the conviction that all reactions in the cell are catalyzed and directed by enzymes, the first commandment commands that enzymology can be relied on to clarify a biologic question. Chemists once bridled at this. But time and again, spontaneous reactions, such as the melting

of DNA and the folding of proteins, are found to be driven and directed by enzymes ------------------------------. The first and crucial step is to find a way to observe the phenomenon of interest in a cell-free system. Should that succeed, one should be able to reduce the event to its molecular components by enzyme fractionation.

2. ***Trust the Universality of Biochemistry and the Power of Microbiology*** The universality of biochemistry from microbes to humans in basic metabolic and biosynthetic pathways has led to the silly quip: "What's true for *E. coli* is true for elephants, and what's not true for *E. coli* is not true." My faith in this universality encouraged me to focus on how prokaryotes, particularly *Escherichia coli*, replicate their own genomes and those of their phages and plasmids. Microbial generation times, unlike those of eukaryotes, are measured in minutes rather than hours and days. This is where the light on replication shines brightest. I made the choice to work with prokaryotes with the confidence that these systems would be reliable prototypes for how the so-called higher organisms replicate their DNA.

3. ***Do Not Believe Something Because You Can Explain It*** In 1954, we observed an activity in an extract of *E. coli* that incorporated the label of [α-^{32}P]ATP into an acid-insoluble form that we presumed to be RNA. While making progress in purifying this activity, we learned of a discovery in the laboratory of Severo Ochoa in New York. While observing an exchange of [^{32}P]orthophosphate with ADP in extracts of *Azotobacter vinelandii*, they discovered an enzyme that converted the ADP and other NDPs into an RNA-like polymer.

Acting on this information, we substituted [α-^{32}P]ADP for ATP and found that our activity was far greater. Clearly ADP was the preferred substrate over ATP. The *E. coli* enzyme we then purified was the same polynucleotide phosphorylase that the Ochoa laboratory had first identified in *Azotobacter*. As was learned later, the role of the phosphorylase was to degrade RNA rather than effect its synthesis. Had we persisted with ATP as substrate, we would surely have found RNA polymerase, the true synthetic enzyme, a year earlier than we did DNA polymerase and several years before it was discovered in 1961 by the late Sam Weiss.

4. ***Do Not Waste Clean Thinking on Dirty Enzymes*** The late Efraim Racker enunciated this commandment, and I have been one of its ardent disciples. A dramatic example is the discovery of DNA replication. I first observed DNA synthesis in an *E. coli* extract in 1955, when I found that 50 counts out of a million of [^{14}C]thymidine were incorporated into an acid-insoluble form. Those few counts above background seemed real because they were susceptible to DNase. Could we possibly figure out what was going on in so crude a system, let alone in an intact cell?

5. ***Do Not Waste Clean Enzymes on Dirty Substrates*** Three years after the hoopla about the synthesis of a viral DNA by our DNA polymerase, serious questions remained. How is a DNA chain started? How is the accumulating genetic evidence for additional polymerases and other factors needed for replication explained? A cartoon at the time showed the apparatus at a replication fork discreetly obscured by a fig leaf, and there were polemical attacks in *Nature New Biology* that dismissed our DNA polymerase as merely a repair enzyme with little relevance to replication, in essence a "red herring."

    Over the years, we had tried a variety of DNA samples to demonstrate the start of a DNA chain. The results were negative or equivocal. Then it dawned on me that we were violating the fifth commandment. We were using a pure DNA polymerase on a dirty DNA substrate: frayed, gapped, fragmented, denatured and heterogeneous. When we finally switched to the intact, single-stranded, circular DNA of a small bacteriophage, we discovered how a DNA chain is started: priming with RNA.

6. ***Depend on Viruses to Open Windows*** Single-stranded phages provided not only the DNA substrate with which we could discover the RNA priming of new chains, but also the enzyme systems responsible for the priming and subsequent replication. The filamentous phage M13 depends on the host RNA polymerase to make a short transcript of an origin region, whereas the icosahedral phage ϕX174 appropriates a complex primosome, used by the host to prime the start of chains on the lagging strand at the replication fork. Whereas the conversion of the single M13 viral strand to the duplex replicative

form was readily resolved and reconstituted, the conversion of the φX174 single-stranded circle was far more complex and required the discovery of 15 new proteins, which constitute the apparatus at the host chromosomal replicating fork. With these many proteins in hand we could attempt to discover how replication was initiated at the origin of the intact *E. coli* chromosome.

7. ***Correct For Extract Dilution With Molecular Crowding*** Cells are gels. Half of the cell dry weight is made up of proteins packed in highly organized communities. That some of their functions, individually and collectively, can be observed despite great dilution (20-fold or more) is a fortunate break for biochemistry. But there is an absolute need in some cases to restore the crowded molecular state, as we learned from our attempts to observe initiation of replication at the origin of an intact chromosome.

    We were given a 5-kb plasmid containing the origin of the 4,000-kb *E. coli* chromosome that is replicated in the cell with the physiological and genetic features of the host chromosome, in effect a minichromosome. When Seichi Yasuda came from Japan with this *oriC* plasmid, I thought we would soon resolve and reconstitute its replication much as we had done with phage φX174. But it took 10 man-years of utter frustration before we finally succeeded in making a cell-free system work.

8. ***Respect The Personality of DNA*** For years, DNA was regarded as a rigid rod devoid of personality and plasticity. Only upon heating did DNA change shape, melting into a random coil of its single strands. Then we came to realize that the shape of DNA is dynamic in ways essential for its multiple functions. Chromosome organization, replication, transcription, recombination, and repair have revealed that DNA can bend, twist, and writhe, can be knotted, catenated, and chromosome organization, replication, transcription, recombination, and repair have revealed that DNA can bend, twist, and writhe, can be knotted, catenated, and supercoiled (positive and negative), can be in A, B, and Z helical forms, and can breathe.

    Especially noteworthy is breathing, the transient thermodynamic-driven opening (melting) of the duplex that facilitates the binding of specific proteins such as the helicase responsible for priming and the onset of replication. Certain DNA sequences are also predisposed to a more extensive form of melting ("heavy breathing") that creates a relatively large opening for transcription. The resulting RNA-DNA duplex (R-loop) can activate an inert origin of replication by altering its structure, even hundreds of base pairs away, which facilitates its opening by origin-binding and replication proteins. Negative supercoiling supplies the energy for the breathing and other features that direct the shape and movements of DNA at the *oriC* origin of replication. These DNA responses have led to an appreciation of the role of transcriptional activation of replication origins near primers in large chromosomes, both prokaryotic and eukaryotic.

9. ***Use Reverse Genetics and Genomics*** Direct genetics, in which a randomly mutated gene can ultimately be linked to a deficiency in a single enzyme, was a landmark discovery in biologic science. This approach served well by providing *E. coli* mutants defective in replication, some in initiation of a chromosome (e.g., *dnaA*) and others in elongation (e.g., *dnaB*, *dnaC*, *dnaE*, and *dnaG*). But randomly generated mutants do not readily disclose the products of their genes nor their particular functions. Nevertheless, these replication mutants were crucial in validating our assays because DNA synthesis was absent in the extracts of mutant cells and restored when extracts or purified fractions from wild-type cells were added. Reverse genetics and genomics have now made the enzymologic approach even more powerful. Unlike direct genetics, enzymology starts with a defined function, after which finding the responsible genes has become relatively easy.

10. ***Employ Enzymes As Unique Reagents*** Biochemistry is replete with examples in which enzymes have been employed as analytic and preparative reagents. From basic research to industrial processes, proteases, amylases, phospholipases, kinases, and phosphatases, etc., have been crucial in operations that were beyond the capabilities of available chemical technology. I will mention just a few examples of applications to DNA and its replication and one from my recent research on inorganic polyphosphate (poly P). The key discovery in

1944 that identified DNA as the genetic substance was based on the destruction by crystalline pancreatic DNase of the factor that transformed one strain of *Pneumococcus* sp. to another. It was the action of this DNase again, as mentioned in commandment IV, which in 1955 made me believe that the few counts of [^{14}C]thymidine incorporated by an *E. coli* extract into an acid-insoluble form signaled the synthesis of DNA. Many more examples can be cited in which an enzyme reagent was decisive: the circularization of linear DNAs by ligases, the creation of "sticky" tails by specific exonucleases used to prepare the first recombinant DNAs, the innumerable uses of restriction nucleases, and on and on.

Possessed as he was with a keen sense of humor, Kornberg concluded his list of commandments with a reference to the movie *History of the World, Part 1*, in which Moses (played by the comedian Mel Brooks) descends from Mt. Sinai carrying *three* tablets enunciating more than Ten Commandments. He drops one and it shatters. He sighs: "Oh well, Ten Commandments are enough!" "Not so," Kornberg entertainingly opined! "The most important of the Commandments was on the lost tablet: 'thou shalt support basic research.'"[33]

Three years later Kornberg was invited by the journal *Trends in Biochemical Sciences* (TIBS) to submit an updated version of the commandments. "When TIBS invited me to submit my current version of the commandments, I realized that after only three years the list needed to be judged again to qualify as my 'top ten'", he stated. "The list is personal, and has been narrowed by my previous focus on DNA replication and my present one on inorganic phosphate," a topic in which prior to his death Kornberg developed a renewed interest (addressed in the ensuing chapter). In the interest of brevity the amended version of Arthur Kornberg's Ten Commandments is not presented here. Interested readers should consult Ref. 34.

It is surely no exaggeration to state that Arthur Kornberg's contributions to our understanding of DNA replication at a fundamental level must be ranked among the great offerings to biochemistry and molecular biology in general and enzymology in particular. To say the least, aside from the

216 *Emperor of Enzymes*

Kornberg on the tennis court.

stellar observations and insights from his own laboratory, his work opened the field of DNA replication to legions of other laboratories around the world. These impressive gains notwithstanding, in the final analysis Kornberg sagely recognized that the book of DNA replication was not ready to be closed. "The awesome and complex assemblies of proteins that start and extend DNA chains will consume the efforts of scientists for decades to come," he wrote.

> Even when we identify each of the many parts of these machines, locate and clone their genes, amplify these proteins by genetic engineering,

and then put an operating humpty-dumpty together again, we will still be short of what we need to know about the replication process.[35]

In 1997 Kornberg told interviewer Sally Hughes that once when he presented an invited lecture in Japan he used the title "DNA Replication From Start to Finish," with the intention of relating current and exciting work on how replication of the chromosome is completed; how it terminates; how it is finished. "I was greeted at the airport by a former postdoc, Kazuhisa Sekimizo," he related.

> Kazuhisa seemed disturbed about the seminar I'd be giving. "What's the trouble, Kazu?" I asked him. He responded, "Well, it's the title of your seminar." "What's wrong with it," I enquired. "I'm going to talk about this exciting work that Jung Lee is doing on termination." He said: "Yes, but when my professor saw the title of your talk, he said, 'You must stop working on DNA replication because Dr. Kornberg says it is all finished!'"[36]

****

In 1969, 10 years after moving to Stanford, Kornberg resigned the chairmanship of the biochemistry department. Having invested a full decade in building and maintaining what was considered by many to be the most outstanding biochemistry department in the country — if not the world, Kornberg was certainly entitled to a break from the chairman's office — an abode described by a former graduate student as

> small, with little decoration or fancy furniture. ---------- It had a tiny table with three legs and three chairs that he used for discussing data with the students. The door was unadorned and there was no nametag. From the hall, it easily could have been the janitor's closet.[37]

Kornberg's preference for a small and unpretentious office was a calculated decision. Noting that it was traditional that chairpersons of academic departments across the world occupy impressively spacious offices, he told interviewer Sally Smith Hughes:

> at Harvard it would be palatial; at Washington University it was also very large. At Stanford I converted the chairman's office to a library.

> I had my office way down the hall so that I would not be bothered or tempted to get involved in administrative affairs. I had a very small office as part of the lab, with a partition on one side. ----------- I placed my office adjacent to my lab and remote from the executive offices.

Following Kornberg's retirement from the chairmanship the position was rotated among the senior faculty in the Department of Biochemistry for five-year terms, beginning with Paul Berg. "It was quite natural that Paul would succeed me as chairman because he is a very interactive person," Kornberg stated. "He is very generous with his time and concern, bright and accomplished, and he carried on what I had been doing."[38]

Kornberg offered no apologies for his concerted efforts to remain engaged with his research to the maximum extent possible while coping with the burdens of a department chair. "I have rather different, and I would say heretical views about administration and teaching," he related.

> I didn't spend more than 10 percent of my time on departmental affairs. I conscientiously asked others on the faculty to manage various issues, so that I was not the custodian of all the physical and functional attributes of the department.[39]

After 10 years in the chairman's office Kornberg could likely rationalize more than a few cogent reasons for deserting the place. But this event was likely to a significant extent precipitated by a position he adopted concerning the admission of minority students to Stanford Medical School. The mid- to late 1960s surfaced a time of considerable social unrest in the United States centering around the war in Vietnam and civil rights, a time that Kornberg characterized as a "frightening period in academic life."[40] At Stanford, one of many campuses around the country where civil unrest seethed, the research and teaching laboratories of the Electrical Engineering Department were occupied and disrupted by students and others, with the encouragement of some faculty members.[41] The atmosphere of racial tension that embraced this period was in no way alleviated by the recognition in many academic institutions in the country, especially in medical schools, that a significant number of racial minority students had a difficult time coping with the curriculum — for which many were poorly prepared. At Stanford this issue led to a decision by the Medical School Faculty Senate to ease admission requirements for minority students.

Not everyone was in favor of this sort of reconciliation, Kornberg prominently among them. As a matter of principle he was opposed to modifying admission requirements to the benefit of students of any racial group. Incensed about Kornberg's position, members of the Stanford Black Student Union released a lengthy and provocative document stating (among other things):

> We don't want brothers in the Med School so they can rap with rich, white heart-transplanting doctors in a few years. No, we need doctors in black communities this morning. If to educate black students at Stanford is to be at the expense of farmers and ham-heads like Kornberg, so be it. ---------------------------------- Why not just pretend Kornberg doesn't exist? We can't do this, firstly because all black medical students must take Kornberg's course and would be in the absurd position of proving their academic worth to this clown; secondly because presumably his influence could be pervasive enough to impede a black student's path to his MD by prejudicing his colleagues: but mostly because we will not tolerate a cracker, institutional-bureaucratic, naïve, slobbering hog to walk around polluting this already purifying campus with his bacon breath.[42]

Kornberg and other faculty at the medical school met with the Black Student Union on several occasions. At one notably contentious meeting angry students quoted Kornberg as stating that he was opposed to admitting more black students because "they are incapable of passing my course and therefore could not pass through medical school here. The three black students here already are doing poorly and we can't afford any more."[43]

Kornberg ended his autobiography *For the Love of Enzymes: The Odyssey of a Biochemist* with a chapter entitled *Reflections of My Life in Science*, in which among other sentiments he voiced some of his feelings about this unpleasant episode. He was clearly hurt by the sordid event, and he bemoaned the fact that he had been badly misjudged and was unsupported by the faculty at large.

> I was accused of racism in the student newspaper and physically threatened by the black students' organization. Instead of rallying to support me some of my faculty colleagues whispered their disapproval of my "insensitivity." How could they have been so insensitive? They must have known about my own struggle about prejudice and my long record of compassion with victims of social injustice and devotion to liberal causes. In the mood of the moment, they abandoned their allegiance to scholarship in their passion for righting an ancient and ugly wrong.[44]

Nor was Kornberg reticent to comment on the outcomes of bowing to student pressure. "The consequences of this general erosion of academic values were predictable," he wrote.

> A student-dominated committee on admissions, directed to see that 20 percent of the class were of black and Hispanic origins, went farther and discriminated against students whose good scholastic qualifications could be attributed to an "advantaged" background. Under these pressures, grades were eliminated and difficult courses were made elective. These attitudes permeated the student body and favored appointments of faculty with a similar lack of scholarly devotion. After a decade of this "cultural revolution" came a gradual return to the sane recognition that scientific rigor is the ultimate basis of medical progress, good clinical practice, and physicianly responsibility.[45]

Paul Berg related his own views about the events surrounding this issue.

> We were in a stage in which we in the medical school believed that it was time to increase the racial diversity of the medical students. In around 1964 there was hardly a black face to be seen among the medical student population. So there was discussion about trying to rectify that. We began recruiting more vigorously and probably gave some preferential treatment to people who were not really qualified enough. From the very get go Arthur did not think that we were doing the right thing.

Everyone made the case that there were excellent undergraduates out there. We had to find them, but also take extenuating circumstances into account. So it was essentially an affirmative action program. This was discussed at a meeting of the Medical School Senate and he likely stood up and made some inappropriate comments. I don't know what he said exactly. Knowing him as well as I did I'm pretty sure that he knew that some of his comments were inappropriate the instant he uttered them. I remember that his wife Sylvy and some of his friends were frankly shocked by what he said at the senate meeting(s). The word got around that he had a racial attitude; this for a guy who was rabid about anti-Semitism! And recognizing this he retired from the debate and eventually the entire issue evaporated.[46]

References

1. Kornberg A. (1991) *For the Love of Enzymes: The Odyssey of a Biochemist.* Harvard University Press, p. 227.
2. Ibid.
3. Ibid., p. 228.
4. Brutlag D, Schekman R, Kornberg A. (1971) A Possible Role for RNA polymerase in the initiation of M13 DNA Synthesis. *Proc Natl Acad Sci* **68**: 2826–2829.
5. Ibid.
6. Letter from William Dove to Arthur Kornberg, Sept. 30, 1971. Reproduced with permission from the Department of Special Collections, Stanford University Libraries.
7. Letter from Arthur Kornberg to William Dove, Oct. 15, 1971. Reproduced with permission from the Department of Special Collections, Stanford University Libraries.
8. Letter from William Dove to Arthur Kornberg, Oct. 18, 1971. Reproduced with permission from the Department of Special Collections, Stanford University Libraries.
9. Kornberg A. (1976) RNA Priming of DNA Replication, in *RNA Polymerase*, Cold Spring Harbor Laboratory Press, 331–352.
10. Kornberg A. (1991) *For the Love of Enzymes: The Odyssey of a Biochemist.* Harvard University Press, pp. 224, 225.
11. Ibid., p. 223.

12. Ibid.
13. Ibid.
14. Letter from Arthur Kornberg to S. Asakura, Aug. 8, 1975. Reproduced with permission from the Department of Special Collections, Stanford University Libraries.
15. Obituary, Reiji Okazaki, Molecular Biologist. *Tanpakushitsu Kakasan Koso* **1**: 874–875, 1976.
16. Kornberg A. (1991) *For the Love of Enzymes: The Odyssey of a Biochemist.* Harvard University Press, p. 231, 232.
17. Ibid, p. 236.
18. Ibid, p. 256.
19. Ibid.
20. Ibid.
21. Ibid., p. 231.
22. Kornberg A. (1976) From DPN to DNA and Back. Unpublished essay delivered at symposium of honor of Carl Cori, Oct. 12, 1976.
23. Kornberg A. (1991) *For the Love of Enzymes: The Odyssey of a Biochemist.* Harvard University Press, p. 241.
24. Ibid.
25. Ibid., p. 249.
26. Ibid, pp. 249–251.
27. Ibid., p. 251.
28. Ibid., p. 228.
29. Roger Kornberg, personal communication, 2014.
30. Deamer DW. (2010) From "Banghasomes" to Liposomes: A Memoir of Alec Bangham, 1921–2010. *The FASEB Journal.* [http://www.fasebj.org/content/24/5/1308.full]
31. Kornberg A. (1991) *For the Love of Enzymes: The Odyssey of a Biochemist.* Harvard University Press, pp. 281–219.
32. Kornberg A. (2000) Ten Commandments: Lessons from the Enzymology of DNA Replication. *J Bact* **182**: 3613–3618.
33. Ibid.
34. Kornberg A. (2003) Ten Commandments of Enzymology, Amended. *TIBS* **28**: 515–517.
35. Kornberg A. (1991) *For the Love of Enzymes: The Odyssey of a Biochemist.* Harvard University Press, p. 239.

36. AK Interview with Sally Smith Hughes, Program in the History of the Biosciences and Biotechnology, Biochemistry at Stanford, Biotechnology at DNAX, Arthur Kornberg, 1997, p. 81.
37. Fuller RS. (2008) A Tribute to Arthur Kornberg 1918–2007. *Nature Struct & Mol Biol* **15**: 2–17.
38. AK Interview with Sally Smith Hughes, Program in the History of the Biosciences and Biotechnology, Biochemistry at Stanford, Biotechnology at DNAX, Arthur Kornberg, 1997, p. 45.
39. Ibid.
40. Kornberg A. (1991) *For the Love of Enzymes: The Odyssey of a Biochemist*. Harvard University Press, p. 309.
41. Ibid.
42. Undated document from the Stanford University Black Student Union.
43. Ibid.
44. Kornberg A. (1991) *For the Love of Enzymes: The Odyssey of a Biochemist*. Harvard University Press, p. 309.
45. Ibid.
46. ECF interview with Paul Berg, Nov. 2013.

CHAPTER THIRTEEN

Polyphosphate

In the mid-1950s, while still at Washington University, Kornberg and his wife Sylvy became interested in a compound called polyphosphate (PolyP): linear polymers containing a few to several hundred residues of orthophosphate linked by energy-rich phosphoanhydride bonds.[1] PolyP is ubiquitous in Nature, but its function(s) is (are) poorly understood. When asked by Sally Smith Hughes what prompted him go off in the direction of polyphosphate, Kornberg related that he once gave a seminar at the University of Wisconsin, and as is standard during such visits, he was shuttled around to various scientists who wished to meet with him.

> I entered somebody's office in which I saw on the blackboard this statement: "seminar today by Arthur Kornberg at 4 PM — on you know what." It crossed my mind that some day I'd like to talk about something that they don't expect me to talk about. There was the wish to do something different from what I'd been doing for forty years.[2]

"There was nostalgia and sentiment centered about working on polyphosphate," Kornberg stated.

> My late wife Sylvy and I had found an enzyme that made a substance that was very much in people's attention then and very baffling. It was this polyphosphate chain. At that time ATP and anything that had high-energy bonds were very much in people's minds. Inorganic polyphosphate, with many hundreds of phosphates linked by such high-energy bonds was a molecule that people knew existed in cells, probably existed prebiotically; very early in evolution of molecules on earth. It was called

a molecular fossil because it had been around from the beginning of time and nobody knew what it did — presumably nothing![13]

Though chemists long regarded polyphosphate as a remnant from earlier evolutionary stages, PolyP is known to have regulatory roles and to occur in representatives of all kingdoms of living organisms, participating in metabolic correction and control of both genetic and enzymatic levels. PolyP is directly involved in the switching-over of the genetic program characteristic of the exponential growth stage of bacteria to the program of cell survival under stationary conditions, "life in the slow lane," as Kornberg liked to refer to that state of bacterial existence.

PolyP participates in many regulatory mechanisms in bacteria. It participates in the induction of rpoS, an RNA-polymerase subunit responsible for the expression of a large group of genes involved in adjustments to the stationary growth phase, and to many stressful agents. PolyP also supports cell motility, the formation of biofilms and bacterial virulence, and it participates in the regulation of levels of guanosine 5'-diphosphate 3'-diphosphate (ppGpp), a second messenger in bacterial cells. Then too, PolyP has a role in the formation of channels across living cell membranes involved in the transport processes in a variety of organisms.[4,5] An important function of the compound in microorganisms — prokaryotes and the lower eukaryotes — is to handle changing environmental conditions by providing phosphate and energy reserves. Polyphosphates are present in animal cells and there is evidence for their participation in regulatory processes during development, cellular proliferation and differentiation.[6]

A potent phosphorylating agent, PolyP is a plausible source of energy and phosphate in the early evolution of our phosphate world. As a potent chelator of metal ions it may be involved in a variety of reactions that exploit divalent cations, including those that help to maintain ATP levels in cells. And, as a metabolic regulator, the compound has been suggested to play a role in the homeostasis of inorganic phosphate, metals and essential nutrients in the face of environmental stresses. In humans polyphosphates have been shown to play a key function in blood coagulation.[7]

In the mid-1990s, after many decades of research on DNA replication, Kornberg and his colleagues published over 20 original research papers and several reviews on polyphosphates. His final literary contribution, a review article written in the fall of 2007, shortly before his death in late October of that year, was entitled "Abundant Microbial Inorganic Polyphosphate, PolyP Kinase Are Underappreciated." In his summary Kornberg pointed out several features of polyP addressed in his own research efforts — and those of others:

— "Inorganic polyphosphate molecules and the polyP kinase enzymes PPK1 and PPK2 that make it are widely distributed among microorganisms and appear to be virulence factors for many pathogens."
— "Several bacterial species with mutations in *ppk1* or its equivalent are defective in growth and survival, motility, quorum sensing (a system of stimuli and response correlated to population density), biofilm formation and in a variety of responses to stress."
— "Because polyP and polyphosphate kinases1 and 2 appear necessary for virulence, they might be targets for the development of novel treatments."
— "Cells of the social slime mold *Dictyostelium disciodeum* produce a bacterial PPK1-like enzyme (DdPPK1) that, if mutated, are defective in development, sporulation, predation and in the late stages of cytokinesis and cell division."
— "DdPPK2, an unusual PolyP kinase from *D. disciodeum* consists of more than 100 globular units, each a tetramer of three distinctive actin-related proteins."

Kornberg challenged the scientific community to devote more interest and energy to this mysterious biomolecule. "Despite many advances, current textbooks and journals make hardly a mention of PolyP," he stated.

> This lack of attention is another instance in which the power of fashion dictates pursuits in science. Not only is polyP often absent from texts on biology and chemistry but, even when noticed, tends to be dismissed as

a "molecular fossil." Biologists should reflect on this attitude and consider how to move polyP from its lonely outpost to a more prominent place along the frontiers of science.[8]

There is a piece written up in *Stanford M.D.* (a medical school periodical) that quotes one of my postdocs coming back from a long weekend and saying to the group: "You know, this weekend I found myself thinking more about the polyphosphate work than I did after a similar weekend when I was working on DNA replication,"

Kornberg related.

And a Chinese postdoc chimed in, "Of course, you behave as if you're trying to get tenure." So that reflects the fact that I do feel more responsibility for the polyphosphate work — responsibility to myself and to my people who have either been hoodwinked or of their own volition are working on a problem that's very much off the beaten path.

"I'm trying to get collaborations or to help other labs work in this area," Kornberg continued.

The work is much more exploratory; it is not done in the depth that I worked on in replication. It's more physiological. But it is our responsibility to show why polyphosphate is important. We didn't have to do that with DNA replication. Anything we did on the enzymes that made DNA was already noteworthy and attention getting. People ask, "What is polyphosphate? Never heard of it; don't care about it." So the only way to make people care is to say: "It's important in some disease — say meningitis."[9]

The contribution we've made that I'm pleased with is in the aging of *E. coli* — meaning when the *coli* culture stops growing because it has run out of nutrients or has toxic elements. So why does it stop growing? Well, that's another matter. But now we find out that when we create mutants that don't have polyphosphate they die in this stationary phase; — call it aging in *E. coli*. How does it cope with stresses and nutritive deficiencies and toxins? All I can say now is that polyphosphate is an important part of the network of responses to stresses. We're trying to get down to the molecular details of how it does that. I think that's exciting. But it doesn't persuade people to drop what they're doing and work on polyphosphate — yet![10]

As interested as he was in the biology and biochemistry of PolyP at a late stage of his career, by 1997 Kornberg was realistic about the fact that the topic had none of the glamour and excitement that DNA replication had held for graduate students and postdoctoral fellows for so many years. "I've had trouble about attracting people eager to work on polyphosphate," he ruefully declared.

> The fact is I'm seventy-nine years old. Why would anyone work with somebody that old, with limited energy and a limited future? I'd rather work with my sons Roger or Tom. They're at the height of their productivity and their careers![11]

"It is illustrative of the problem of doing something that is pioneering, or unorthodox, or novel," Kornberg concluded.

> The climate in science has been, and is even more so now, unreceptive to things that are labeled courageous, but could be foolish, unproductive efforts. One of the great flaws in our granting system is that it is poor in support and sympathy for doing something that might turn out to be unproductive.[12]

"Arthur never passed on an opportunity to explain science and his work," his son Tom Kornberg related.

> Everyone who came in contact with him during the last decade of his life learned about the importance of PolyP. He regaled his dinner partners at the Nobel banquet in 2006 with the topic and even inculcated his grandchildren. It's no accident that my son Zac (a college student) is working on polyphosphate as we speak![13]

Kornberg would likely have appreciated two contributions to the literature about polyphosphate that surfaced at the time of this writing, years after he passed on; one entitled "Signaling Properties of Inorganic Polyphosphate in the Mammalian Brain," published by Kira

Holmstrom *et al.* in *Nature Communications* in January 2013, and another by Michael J. Gray *et al.* entitled "Polyphosphate is a Primordial Chaparone," published in the March 2105 issue of *Molecular Cell.*

References

1. Phosphoric Acids and Phosphates. http://en.wikipedia.org/wiki/Phosphoric_acids_and_phosphates.
2. AK Interview with Sally Smith Hughes, Program in the History of the Biosciences and Biotechnology, Biochemistry at Stanford, Biotechnology at DNAX, Arthur Kornberg, 1997, p. 152.
3. Ibid.
4. Kornberg A. (1995) Inorganic Polyphosphate: Toward Making a Forgotten Polymer Unforgettable. *J Bact* **177**: 491–496.
5. Brown MRW, Kornberg A. (2004) Inorganic Polyphosphate in the Origin and Survival of Species. *Proc Natl Acad Sci USA* **101**: 16085–16087.
6. Phosphoric Acids and Phosphates, http://en.wikipedia.org/wiki/Phosphoric_acids_and_phosphates.
7. Ibid.
8. Kornberg A. (2008) Abundant Microbial Inorganic polyphosphate, Poly P Kinase Are Underappreciated. *Microbe* **3:** 119.
9. AK Interview with Sally Smith Hughes, Program in the History of the Biosciences and Biotechnology, Biochemistry at Stanford, Biotechnology at DNAX, Arthur Kornberg, 1997, p. 82.
10. Ibid.
11. Ibid.
12. Ibid.
13. ECF interview with Tom Kornberg, May, 2014.

CHAPTER FOURTEEN

A Life of Writing

Arthur Kornberg was an articulate and prodigious writer, an activity he clearly enjoyed, as evidenced by his prolific and enduring output of written material. The 463 publications listed in his curriculum vitae include multiple essays, editorials and commentaries on various aspects of basic biomedical research (see Appendix). He prepared thoroughly for invited lectures and talks, frequently making comprehensive written drafts in anticipation, and typically completed such writing well ahead of time. His son Roger recalls that when his father once gave a named lecture in the United Kingdom that required a manuscript scheduled for publication many months later Kornberg delivered the completed paper at the time of the lecture. The lecture organizers were so astounded at such an early delivery that they asked him to take it back, pleading that they were totally unable to deal with it at the time of the lecture for fear that they may lose track of it![1]

Kornberg's writing embraced a variety of genres. Beginning in 1970 he maintained a collection of notebooks — diaries of a sort. At the time of his death a total of 125 notebook-size diaries were stored in his Portola Valley home. Bereft of personal reflections, philosophical musings and reminiscences that grace the pages of more traditional diaries, the entries mainly document events related to his professional and non-confidential aspects of his personal life. Strictly chronologically ordered, they include a *potpourri* of schedules for meetings, invited seminars, notes on seminars by others, a travelogue of places he visited, people he met and for what purpose, menus for special dinners to which he was invited — and more. In addition to entries rendered in his tiny but immaculate handwriting the diaries are crammed with illustrative photographs and pamphlets neatly and sometimes elaborately trimmed or folded to accommodate the size of the notebook pages. One is left to wonder where he found the time to devote to this clearly labor-intensive effort.

A randomly selected but by no means isolated example of the eclectic entries that filled Kornberg's notebooks is represented in his meticulous documentation of a trip with his family to the Jackson Hole and Yellowstone National Park areas in mid-July 1994. His record of the trip begins with pasted air tickets underscored with the comments:

> Flight delay, switched to later flight. ---------------- 10' drive to Jackson. Main streets congested with traffic. Small town nestled in mountains — attractive despite few shops and large crowds. Wooden sidewalks, eaves and balconies, Western décor without being offensive. Dinner at Bubba's barbeque. Excellent! Superb baby back ribs, chili, fries, pies. Very friendly service.

Photos interrupt these notes, some original, a good number taken from brochures and magazines. A map of Yellowstone Park occupies an entire page and is marked to identify regions of the park visited and the animals viewed. The page immediately following this entry reveals notes on a seminar he heard on "Regulation of Cellular Calcium by Yeast Vacuoles." This is followed by notes from a presentation by one of his students at the weekly Kornberg Journal Club, while the subsequent entry features a pasted itinerary of the XOMA (a biotechnology company) Board of Directors Meeting, held in Santa Monica, California. Each book begins with an alphabetized list of names of all individuals mentioned in the particular volume, including the relevant page numbers.

"When we were children Arthur wanted us to keep diaries," Roger Kornberg recalled. "He would insist on us writing in them each day and he would edit our spelling and grammar. The idea of keeping a diary appealed to him greatly. He loved to do it. He loved to write — a skill reflected in most if not all his writing."[2]

Many among the imposing collection of essays, editorials and tributes that Kornberg contributed to the scientific literature reflect his views on various aspects of science and the scientific culture in general, as well as personal reflections on various phases of his own career. He also emphatically deployed this format to decry the shortage of federal funding for basic biomedical research in the United States, a topic discussed more fully in Chapter 15. Many of these essays had their primary origin in invited talks he presented at scientific meetings, symposia and other forums. Their breadth is reflected in the evocative titles that many of them bore:

The Two Cultures: Chemistry and Biology
The Recent Revolution in Biology
Future is Invented, Not Predicted
The Support of Science
Research, the Lifeline of Medicine
Of Serendipity and Science
Basic Motives of a Professional Life
From DPN to DNA and Back
For the Love of Enzymes
Ten Commandments: Lessons from the Enzymology of DNA Replication

In "The Two Cultures: Chemistry and Biology," the topic of a lecture delivered at a meeting of the American Association for the Advancement of Science in February 1987 and published later that year, Kornberg traced the historical roots of chemistry and biology.[3] "Arthur was always keenly aware of the differences between chemistry and biology, which were very notable at the time," Roger Kornberg related. "He really was a chemist at heart, but a biologist in practice. He came from the great tradition of biochemistry, which of course descended from chemistry. He was also in every way a chemist in his passion for numbers and for precision, and for chemical transactions."[4]

"The basic medical sciences, which in my school days were completely discrete from one another, have now effectively been merged into a single discipline," Kornberg wrote, referring to the growing discipline of molecular biology, an intellectual union that as mentioned previously, he was sometimes skeptical about, especially concerning what he perceived as the parallel decline of the discipline of enzymology, a topic of investigation that he dearly loved above all others. He posed the rhetorical question:

> Where has the development of this new branch of biochemistry, called molecular biology, fallen short? In its rapid and turbulent growth molecular biology has washed away much of the bridge to chemistry. In the rush and excitement over the new mastery over DNA, attention in biochemistry departments has been sharply shifted to major biological problems of cell growth and development and away from chemistry. Training in enzymology and its practice have been neglected. Most biochemistry and molecular biology students are introduced to enzymes as commercial reagents and treat them as faceless as buffers and salts. As long as this inattention to enzyme chemistry and basic biochemistry persists, the fundamental issues of cell growth and development will not be resolved and their application to degenerative diseases and aging will be delayed.

"Molecular biology falters when it ignores the chemistry of the products of the DNA blueprint — the enzymes and proteins, and their products — the integrated machinery and framework of the cell, he wrote. "Molecular biology appears to have broken into the bank of cellular chemistry, but for lack of chemical tools and training, it is still fumbling to unlock the major vaults."[5]

When interviewed in 1997, he told Sally Smith Hughes:

> Biochemistry departments by and large have veered very sharply in the direction of biology. They study very complex phenomena, like cancer, aging, development and differentiation — exceedingly difficult biological questions. Doing so, they leave the previous domains of biochemistry to the chemists, who take it up to some extent but with their own cultural bias. They look with sharper focus at the chemical structures and try to improve on them. It is an arrogance that may be justified by some results, but to improve on a chemistry that has taken two to three billion years to evolve is difficult to do. Chemists think they can make a better enzyme and understand it well enough to endow it with new properties. Take catalytic antibodies. Antibodies are highly specific. Could they be engineered ------ to do things that enzymes have not been found to do?[6]

Kornberg opined that at least two driving forces had moved biochemists away from addressing "simpler problems."

> One driving force is the possibility generated by genetics and new techniques to apply chemistry to biological problems that were previously beyond reach. Animals and cells can be engineered to lack a function or to overproduce a function. ––––––––––––––––– The second reason is fashion. There are always pressures —— economic, social, other pressures — to do something meaningful — obtaining a grant, getting recognition from the community and peers, doing something of direct human relevance. The focus on humans has been and is even more so now of greater importance than studies of rats or bacteria. To do the reverse would seem counterintuitive. Yet we know historically and can prove in case after case that solving difficult human biologic problems has depended on the unanticipated extension of basic scientific enquiries that have nothing to do with ultimate application to medicine. ––––––––––––––––––– Biochemistry departments have been renamed Biochemistry & Molecular Biology. There is a shift away from the traditional roots in organic chemistry and intermediary metabolism. Metabolism has become a dirty word in biochemistry. It used to be its core. *E. coli*, the guinea pig of early molecular biology, is now a four-letter word to funding agencies and the academic community.[7]

Kornberg received numerous invitations to share his views about chemistry and biology in public, especially concerning the origin and growth of these disciplines. He never wavered in his defense of enzymology. In an essay entitled "DNA Replication," based on a lecture he presented at a conference convened in Italy in 1988, he referred to the conventional tactic of understanding cell biology as the classical approach in which

> one struggles to observe a cellular function in a cell-free system, find the responsible enzymes by fractionation of cell extracts, and then purifies the enzymes to homogeneity. –––––––––––– The reverse approach, call it neoclassical, has been especially popular in recent decades. In this approach, one first obtains a structure and then looks for its

function. The protein — a hormone, a receptor, an induced polypeptide — is amplified by having cloned the gene and is preferably small and stable and commercially available. By intensive study of the protein and homologous proteins one hopes to get some clues to how it functions. In my work on nucleic acid synthesis I have followed the classical route.[8]

As mentioned various times in earlier chapters, as wedded as he was to enzymology, Kornberg nevertheless, somewhat paradoxically, expressed unreserved appreciation for the contributions of molecular biology as that discipline emerged and matured. Indeed, he added the term *gene hunters* to his lexicon of *microbe hunters, vitamin hunters* and *enzyme hunters* that he liked to deploy. In the opening chapter of his autobiography *For the Love of Enzymes: The Odyssey of a Biochemist,* discussed later in this chapter, he wrote:

> Each age, with its particularly bountiful quarry, was seen as golden. The current age of gene hunting, with its inexhaustible source of genes and a cheap, efficient arsenal to capture them, is incontestably the most golden. There has been no other hunting to match it. Recombinant DNA, the engineering of genes, and related techniques of DNA chemistry constitute what may well be the greatest technological advance in the history of biology and medicine.[9]

"Genetic engineering and related biotechnologies represent the most revolutionary advance in the history of biological science," he told an audience in a talk delivered at a symposium convened at the Rockefeller University in September 1995 and published as an essay entitled "Chemistry — the Lingua Franca of the Medical and Biological Sciences."

> We have an inexhaustible supply of genes and simple and efficient techniques to track and capture them. The term revolutionary is generally overused, but not here. The effects of this advance on medicine, agriculture, and industry can hardly be exaggerated.[10]

A few years later in an essay called "Centenary of the Birth of Modern Biochemistry," Kornberg wrote:

> The extraordinary advances in the cloning of genes and sequencing of genomes provide essential knowledge that will be of increasing and inestimable value. These developments now beg for the biochemistry to understand the functional and organized units of the cell and organism. Just as remarkable are the techniques to knock out genes and alter levels of their expression, which provide profound insights onto physiology and disease.[11]

Kornberg's penchant for the written word yielded six books, three of which were devoted to DNA metabolism in the broadest sense. *Enzymatic Synthesis of DNA*,[12] published in 1961, emerged as a small book contributed to the *CIBA Lectures in Microbial Biochemistry* presented annually at the Institute of Microbiology in New Brunswick, New Jersey. A more substantial contribution entitled *DNA Synthesis*[13] was published by W. H. Freeman & Co in 1974, a time when considerably more was known about DNA replication. In the preface to this volume Kornberg noted that the book emphasized biochemical rather than physiological aspects of DNA synthesis, stating that its scope had been broadened beyond that of *Enzymatic Synthesis of DNA* to include topics pertinent to DNA, notably the precursors of DNA synthesis to include DNA repair, genetic recombination, restriction and transcription. The book was handsomely illustrated by Charlene Levering, a skillful graphic artist at Stanford.

"Writing *DNA Synthesis* ⸺⸺⸺ surprised me in many ways," Kornberg stated years later.

> First, the effort was far greater than I imagined. Very little from the lecture notes and reprints could be lifted and placed in the right context and still remain readable. The required time had to be captured by exploiting a latent senile insomnia, generally from two to six in the morning; these became the most productive and enjoyable hours of my day. I was also surprised by the pleasure I found in reworking and polishing sentences and paragraphs for brevity and clarity, a satisfaction I had never found in crossword puzzles or other word games.[14]

By 1978, four years after the publication of *DNA Synthesis*, Kornberg arrived at the sober realization that "the book needed updating if it was to remain useful to me as well as to others."[15] This time Kornberg and his wife Sylvy spent a month in Italy at the Villa Serboloni, a sumptuous retreat overlooking the confluence of Lake Como and Lake Leccho operated under the auspices of the Rockefeller Foundation in New York. The revised work was published in 1980 and emerged as a volume nearly twice the size of *DNA Synthesis* under the title *DNA Replication*.[16] Kornberg devoted the first half of *DNA Replication* to the enzymology of DNA, a field that in his view had greatly expanded. The rest of the book addresses discussions of DNA supercoiling, binding and twisting enzymes, as well as expanded accounts of DNA ligase, nucleases that attack DNA and inhibitors of DNA replication. He also surveyed the replicative life cycles of bacterial and animal viruses and of plasmids and organelles, notably mitochondria.

The continued progress in the field of DNA metabolism prompted Kornberg to consider writing a supplement to *DNA Replication* with the intention of releasing further supplements as he deemed necessary. But he ultimately abandoned this notion in favor of a second edition of the book. This volume, released in 1992,[17] weighed in at over 900 pages and

Tania Baker.

was co-authored by Tania Baker, a graduate student in Kornberg's laboratory, a collaboration that she described as an "incredible experience."

Baker might not have become a scientist if she hadn't been a babysitter in college. Her employer, geneticist Richard Burgess, was interested in her as a full-time nanny, but thought so highly of her that he gave Tania a job washing dishes in his laboratory at the University of Wisconsin, Madison — the beginning of a career in science that now includes a stellar reputation, including a Howard Hughes Investigatorship.

"We got along well." Baker said.

> I think Arthur liked me. He liked my energy and my spunk — and I'm the sort of person who stands up for herself — right from the beginning, although he and I didn't always agree. When anyone wrote a manuscript, as it shaped up he'd distribute it to the entire lab group for comments. And I tended to comment quite extensively on some of the manuscripts and that got Arthur and me off on a good start. And since it was a difficult time in his life when Sylvy was ill, he gave me several opportunities to speak in his place. I'm sure this helped advance my career. Of course he made sure what I was going to say! The same thing with the book. He told me that he simply didn't have the energy to do it alone. He both wanted someone to help with the writing — and to help keep him on task. When he was writing *For the Love of Enzymes* he would ask me to read sections because he liked the way I edited.[18]

Despite her favored status in the Kornberg laboratory, Tania Baker was also a subject for her share of criticism and reprimand, some of which she revealed in the private publication in Kornberg's memory referred to earlier, that emerged shortly after his death. "Each Monday morning he would stop by my bay and said either; 'How are you? or 'How was your weekend?', which I quickly realized meant: 'What's new with the project?'"[19]

> I was a pretty hardworking student so I usually had data, and we would talk about it for a while. But the on the way out of my lab area he would say: "Now you really should be taking this opportunity of living in the SF Bay Area to better yourself! How about the opera, the symphony

or a Stanford football game?" But I could never win. If I did spend the weekend doing something cultural, the ending speech would be; "Now Tania, where are your priorities", or "this isn't charm school." But I think my favorite Arthur quote is: "Are you planning your life like you are planning to fail?" I believe he said it to me twice, once kindly, once less so, but both were very useful. In each case I was going over plans, a mix of experimental and career path plans. We call them exit strategy talks now, and he was frustrated when I was leaning too much toward a "safe route." .... The most amazing thing about Arthur's lab is that although he was tough on us, he believed in us and in our ability to solve problems, learn new things, and succeed.[20]

"He taught me rigor. Writing is hard, unrelenting work and we certainly had our moments of frustration," Baker commented.

But Arthur showed amazing flexibility and patience in writing with me. I'm dyslexic, and thus spell very badly. In contrast, Arthur spelled well. These were the early days of word processors, so Arthur got to see these glaring errors. He could not fundamentally understand how it could be that I could write, but never learn to spell![21]

The numerous illustrations that adorn the pages of both editions of *DNA Replication* were again the work of Charlene Levering.

In the midst of his intensely busy life as a scientist, administrator, advocate for basic science and writer, Kornberg found the time and energy to render an autobiography entitled *For the Love of Enzymes: The Odyssey of a Biochemist*,[22] mentioned several times earlier in this text. The book was simultaneously released in the US and the UK in 1989 and is dedicated to Sylvy.

For the Love of Enzymes chronicles Kornberg's scientific career, beginning with his assignment to the NIH in 1944, with emphasis on the biochemical journey that led him to the discovery of the first DNA polymerase in the bacterium *E. coli* and the subsequent triumphant assembly of the *E. coli* replication machine — the so-called replisome. A penultimate chapter introduces the reader to the emergence of the discipline of molecular biology and the origins of recombinant DNA

For The Love of Enzymes. Kornberg's autobiographical rendering.

technology and genetic engineering, while a concluding chapter entitled "Reflections on My Life in Science" surfaces some personal themes.

For the Love of Enzymes strikingly reveals Kornberg's authoritative and accomplished writing skill and command of the English language. One's enjoyment of his prose and the story he tells is heightened by a series of informative illustrations artfully rendered by Charlene Levering. Not surprisingly, the book won an award from the American Medical Writers' Association.

Others thought equally well of the book. An unusually eye-catching promotion reads as follows:

> In 1645 the Japanese samurai Musashi Miamoto wrote *A Book of Five Rings*, which described the attitudes necessary for individual success. Though he was a swordsman, his book is not limited to combat

but addressed the much broader question of how to achieve excellence in life through study, discipline, and planning. It is still avidly read in Japan today. Arthur Kornberg's book is a modern-day *Book of Five Rings* that replaces the medium of swordsmanship with that of biochemistry, particularly enzymology. *For the Love of Enzymes* does not describe a single lucky or hard-won accomplishment. Rather, it is the story of thirty years of decisive campaigns, nearly all of which led to insights of major significance. In relating his story, Kornberg never avoids the difficult question of "why": why he felt classical nutritional studies had reached a plateau; why he turned to enzymology as a discipline in which the important answers would be found and why he believes the study of enzymes will grow ever more important as we face the new scientific frontier of brain function.[23]

Kornberg's scientific colleague Joshua Lederberg, who as mentioned earlier served as head of a new Department of Genetics at Stanford University from 1958–1978, opined that

> *For the Love of Enzymes* is not the book that Arthur Kornberg intended. His first drafts were focused on the exposition of the facts of biochemistry he felt should be more widely learned by the general public. His friends and admirers sought to persuade him of the legitimate interest in a biography of a science, in its day-to-day challenges, in the development and personal character of its practitioners, in the interplay and environmental influences that may lead to extraordinary accomplishment. Gradually, almost grudgingly, the successive drafts of his manuscript have responded to that appeal; we now have a work that combines scientific exposition with autobiographical memoir. Kornberg has played such a commanding role in the biochemistry of the gene: it is impossible to tell its history and exclude the personality that brought so much of it about.[24]

Others who expressed opinions about *For the Love of Enzymes* included the Harvard biochemist Eugene Kennedy, who thoughtfully emphasized the logic that led Kornberg to the discovery of a DNA polymerase, "a logic sometimes lost in communicating the history of this class of enzymes" Kennedy wrote:

> While still at [the] NIH he purified an enzyme that split the coenzyme nicotinamide adenine dinucleotide (NAD) at the pyrophosphate bond.

His success confirmed his belief in the power of enzyme purification as a tool in biological research, a doctrine of which he was to become the passionate advocate. ‑‑‑‑‑‑‑‑‑‑ One of the products of the reaction catalyzed by the potato enzyme is nicotinamide ribose phosphate (NRP). The availability of NRP led him to study the synthesis of NAD.

This discovery was an eye-opener. The elimination of inorganic pyrophosphate (P-P) from a nucleoside triphosphate (such as ATP) first observed here proved to be a very general feature of biosynthetic reactions. Because NAD synthesis also involved the transfer of an activated nucleotide building block, it also offered insight into the much more challenging problem of the biosynthesis of DNA and RNA, then looming only distantly.[25]

While fully inclusive of his own role in the science that he presented, *For the Love of Enzymes* offers little information about Kornberg's childhood, youth or family, a reality that as mentioned in the preface strongly motived my own contribution about Kornberg's life. Arthur's son Tom suggests that this was intentional and offers the opinion that his father believed that readers would have little interest in knowing details of his personal life.[26] "I wished to use the chronology of my career only to organize the narrative and introduce personal elements where they might leaven and humanize science," Kornberg wrote in the preface to *For the Love of Enzymes*.[27] This viewpoint is reinforced in a review of the book by then *Nature* editor Peter Campbell, who at the very outset reminds the reader that

> this book was written under rather sad circumstances during the tragedy of the long and unremitting illness of the author's wife. This unhappy disposition alone may account for Kornberg's reluctance to avoid sentimental reminiscences of years gone by, most, if not all, would likely conjure up thoughts and memories of happier times with Sylvy.[28]

Peter Bondy of Yale University expressed disappointment at this omission. When reviewing the book, he wrote:

> Toward the end there is one chapter describing his family, of whom he is very proud. He also mentions his first wife, Sylvy, with great affection. But neither his family nor his co-workers come across as real people. There are hints that important material lies unrevealed. For example,

when Kornberg got the Nobel Prize, Sylvy said: "I was robbed!" Kornberg calls this a "quip", but in the next sentence he sketches quickly some of Sylvy's important contributions to the work. Was she really joking or did she feel that her part in the enterprise had been misunderstood? Unfortunately, Sylvy doesn't come through as a rounded person. Indeed, although she was certainly an important part of Kornberg's life, she seems a mere shadow in the book. His three sons also appear only briefly and are not presented in a fashion which shows them as real people. For example, Tom was apparently a promising cellist, but dropped his musical career when he developed a painful neuroma on the first finger of his left hand. Thereafter he built a career in science. But did he give up his music altogether? A little more information would have made Tom a real person to this reader. ‑‑‑‑‑‑‑‑‑‑‑‑‑‑‑‑‑‑‑‑‑‑‑‑‑‑‑‑‑‑‑‑‑ A good book, and it would have been better if Kornberg and his family and colleagues had been more central. As it is, the detailed description of science overshadows the human aspect of a great career.[29]

****

An attentive and revered wife and mother, as well as a competent biochemist, Sylvy Kornberg was often referred to by her sons as "Ship" rather than Mom. Roger Kornberg explained that

> one summer in the early 1950's we were in Pacific Grove, California where my father took van Neil's famous microbiology course in preparation for teaching the subject himself at Washington University in St. Louis, where he took up the chair in the subject. Tom, Ken and I shared a bedroom in a beautiful Chinese Pagoda situated in a clearing in the forest adjoining Asilomar, which we rented for the summer. One evening our parents came in to say goodnight and young Tom (then a lad about six years of age), doubtless influenced by his fascination with boats or ships he had seen that day, said "Good night Shippy." The name stuck, in part because I recall my mother was amused by it.[30]

By all accounts Sylvy Kornberg was extremely well liked by all who knew her. She is universally described as a charming and pleasing personality. "Sylvy was a marvelous person, and a marvelous scientist at a time when women were not encouraged to be active scientists," Dale Kaiser stated. "She was also a very warm person and very loving to Arthur."[31] Very much the traditional matriarch in the Kornberg family, Sylvy was deeply loved by her three sons. "I know from my relationship with Roger, Ken and Tom, that she was a very special person in their lives," Jim Spudich, Arthur Kornberg's first graduate student at Stanford and a frequent visitor to the Kornberg household, recalled. "I viewed her as someone of obvious importance to the family."[32]

While in St. Louis Sylvy worked in Arthur's Washington University laboratory and occupied a bench directly opposite that of Bob Lehman. "When I first joined Arthur's lab she was working on an enzyme that made polyphosphate, an enzyme that turned up quite by accident," Lehman recalled.

> During some of the early experiments involving DNA polymerase Arthur added ATP to an extract of *E. coli*, which generated polyphosphate. Sylvy pursued that observation and purified polyphosphate kinase, an enzyme that supported the synthesis of polyphosphate. When the DNA polymerase work really started to get going she joined the group working on that hot topic. After we moved to Stanford the time came around when Sylvy no longer had the time to do serious research and she eventually stopped working in the lab.

"Arthur adored Sylvy." Lehman recounted. "She was truly the love of his life."[33]

Sylvy Kornberg was not reticent about informing her husband when she was displeased with him. Lehman relates that when he was a postdoctoral fellow in Kornberg's Washington University laboratory he once had tickets to the St Louis opera, an occasion to which he was very much looking forward. But an experiment that he was engaged in that afternoon (the results of which Kornberg was eagerly awaiting) ran late and to his disappointment he had to forgo the opera. The following morning Sylvy who was then still working in the laboratory asked Lehman whether he had enjoyed the opera. The disconsolate Lehman informed her of the events that had transpired the previous day and noted that Arthur had been eagerly awaiting the outcome of the experiment he was engaged in. "She promptly read Arthur the riot act," Lehman laughingly commented. "Slave driver was one of the more polite phrases she used!"[34]

The illness that plagued the young life of Sylvy Kornberg referred to above by Peter Campbell, presented as corticodentatonigral disease, a rare, progressive and ultimately fatal neurodegenerative disease involving the cerebral cortex and the basal ganglia of the brain, characterized by marked disorders in movement and cognitive dysfunction. Little is known about the etiology of the disease. Eric Shooter, a celebrated Stanford neurobiologist and a friend and neighbor of Arthur and Sylvy Kornberg, offers the opinion that corticodentatonigral disease may belong of a class of neurodegenerative disorders associated with the pathological accumulation of a protein in the brain called tau. Alzheimer's disease, wherein tau protein is deposited within neurons in the form of so-called neurofibrillary tangles, is the best known of these illnesses.[35]

Sylvy began manifesting neurological symptoms and signs sometime in 1980. Her first awareness that something was amiss arose when she began to experience tingling in her fingers while playing the piano.

Sylvy Kornberg in her later years.

Within a year she was totally unable to play the instrument as symptoms spread to her arms and legs. "During this harrowing time Arthur, Sylvy Tom, Ken and I were together almost every evening and absolutely every weekend, offering whatever encouragement we could to my mother and father, during an agonizingly difficult time," Roger Kornberg related.[36]

Kornberg moved heaven and earth in the hope that his wife's condition might be treatable. He travelled all over the country and pulled every string he knew — to no avail. His friend Alex Zaffaroni (about whom more is related in a later chapter) was well connected in the pharmaceutical industry and assembled a team to search for any treatments that might be effective. By 1983 the sad prognosis was clear. Arthur was totally distraught. "He went through the motions of going to work and taking care of routine issues, but he was clearly depressed," Ken Kornberg related. "My family moved to his house in 1984 to help Arthur and Sylvy cope with the deteriorating conditions. Eventually we had around the clock nursing care."[37] Sylvy was ultimately confined to a wheelchair and died at home on June 6, 1986.

Kornberg dedicated his autobiography *For The Love of Enzymes* to Sylvy with the words: *In memory of Sylvy, my great discovery.* In the preface he wrote:

> The main impetus for me to tell this story came from the tragedy of my wife Sylvy's long and unremitting illness and the welcome diversion we found in recalling our adventures in science so inextricably woven into our lives together.[38]

The tragedy of Sylvy's illness and death almost destroyed Kornberg. Those close to him felt and shared in his pain to the extent that anyone could. "Around that time Arthur was literally devastated," Paul Berg related. "He couldn't think. He would sit in his office for hours just staring out the window."[39] "Losing Sylvy was an enormous tragedy for Arthur," Bob Lehman commented. "I lived with him through that horrendous time. Sylvy was the love of his life and for him to watch her die such a lingering death at a relatively young age was agonizing."[40]

Following Sylvy's tragic demise Arthur was essentially alone, both physically and emotionally, and in due course announced his intention to marry Charlene Levering, the graphic artist who illustrated *DNA Synthesis*, the first edition of *DNA Replication* and the second edition of *DNA Replication*, authored with Tania Baker. The couple married in 1988, two years after Kornberg lost Sylvy. When Arthur shared this intention more publically one of his close friends commented that he should consider the decision to marry again the highest compliment he could pay to Sylvy, implying that his marriage to Sylvy had been so marvelously close and supportive that he wanted to reacquire the comfort and companionship of marital status.

"In the end everyone who knew Arthur recognized that he deserved to have someone like Charlene around," Berg stated. "But I don't think that the boys (Roger, Tom and Ken) were especially happy with her. She resented them at times, especially when she moved into Arthur's home. There was a certain amount of tension around that."[41] "Charlene tried to draw Arthur into her personal fold at the expense of his relationship with Ken, Tom and me," Roger Kornberg acknowledged. "The three of us disliked, but tolerated her."[42] Unimaginably, tragedy revisited Kornberg when in 1995, seven years after he married her, Charlene died of leukemia. "Once again he was forced to endure the anguish of losing a loved one," his son Tom commented.[43]

Around that time his friend and colleague Alex Zaffaroni professionally introduced Kornberg to a Palo Alto travel agent, Carolyn Frey Dixon. "One day I got a frantic phone call from Alex, who said: "Carolyn, you must help my very best friend who is in terrible trouble," Carolyn related.

> This was on a Thursday afternoon and the office was frantically busy. I was not taking on any new clients. So I told Alex I'd do my best to help. Next I had a phone call and this beautiful voice said: "Carolyn, you don't know me but this is Arthur Kornberg. Let me tell you what I need." He was in terrible trouble. He'd been using a travel agent who had recently had a nervous breakdown and had not done anything she was supposed to with respect to his immediate travel needs. He was on route to New York and then to Europe — a big speaking trip. No hotel reservations or airline reservations had been made. This was

5:00 PM on a Thursday afternoon and he was due to leave on the upcoming Saturday morning.[44]

I told one of my assistants that we had a situation on our hands and asked her to stay late that evening. It was night in New York and every single hotel I called was booked for some major conference in the city. So I called an old friend who was the manager of a hotel in New York who promised to see what he could do. He called me back and told me that he had found a room. It was an expensive presidential suite. I told him that was fine! Long story short we put the whole thing together and delivered everything to him on a Friday afternoon. Everything worked out perfectly and he called me from New York to tell me so.[45]

In due course Kornberg and Carolyn cultivated a social relationship — and married in December 1998.

****

Two more books issued from Arthur Kornberg's prolific pen before his death in 2007. As discussed in Chapter 17, in the early 1980s Kornberg, together with several colleagues at Stanford and several outside the university, entered the private sector and founded a biotechnology company called the *DNAX Institute of Cellular and Molecular Biology, Inc.* — *DNAX* for short, a remarkable, indeed a surprising venture for someone so deeply rooted in the academic universe and the goals and philosophy of "pure science." His venture into the world of biotechnology prompted Kornberg to offer up a book that he dubbed *The Golden Helix: Inside Biotech Ventures*, published in 1995.[46] "In this narrative, which features the DNAX–Schering–Plough partnership, I describe how it made *DNAX* a world leader in basic immunology and at the same time generated for Schering–Plough multiple candidates for drug development far earlier than expected," Kornberg wrote in the preface.[47] Aside from its rendering of the history of the *DNAX*–Schering–Plough venture *The Golden Helix* focuses on the principal individuals involved, notably Alex Zafferoni, who Kornberg described as "a remarkably successful and charismatic scientist-entrepreneur, whose vision and enthusiasm were instrumental in the founding of the *DNAX Institute of Cellular and Molecular Biology*" (who died in March 2104), and on Robert Luciano, CEO of the Schering–Plough Corporation. Though principally focused on the history of *DNAX*, *The Golden Helix* also profiles several other prominent biotechnology ventures.

The book prompted an unusually scathing review in the *New England Journal of Medicine* in late 1995, one too lengthy to merit reproduction here. But a few paragraphs will give the reader the flavor of its highly critical and in some sections frankly insulting nature.

> *The Golden Helix* could have been torn from a company's brochure for high rollers at the biotech gambling table (or, as Kornberg puts it, "the biotech crapshoot") the critic wrote. "But it is a real book, produced by a legitimate publishing house, and it is about real people. ---------------- *The Golden Helix* portrays an academy strongly motivated by personal gain through commerce. In portraying a campus where hucksterism replaces objectivity, the book reveals the line between professionalism and self-aggrandizement. University presidents, medical school deans, department heads, and others concerned with the way professors should (and do) relate to industry will learn from reading this book.[48]
>
> Kornberg boasts about how he and his academic colleagues played their parts on behalf of *DNAX*. These university professors toured Tokyo, London, Paris, and New York on fund-raising campaigns for a private company. On their way to Japan, he and other prestigious academics stopped in Honolulu to rehearse their slides and presentations — much like graduate students preparing their 10-minute talks for a national scientific meeting, except that the professors enjoyed memorable hospitality ("the French cuisine in Tokyo and Osaka was easily of three-star quality"). Kornberg tells us how he and his traveling companions reveled in "elegant settings in France and Ireland," but fails to mention that their graduate students were working for pittances back home at Stanford and Harvard.[49]

The sharply invidious tone of the review prompted a brisk series of letters from some of Kornberg's colleagues to the *New England Journal of Medicine* that essentially panned the critic, including a brief missive from Kornberg himself, who wrote: "In response to the personal attack by

Robert Schwartz in his review of my book *The Golden Helix: Inside Biotech Ventures,* I simply quote Ezra Pound: 'I can spot a bad critic when he starts by discussing the poet and not the poem.'"[50]

The final book that Kornberg authored has a fascinating and thoroughly charming history. This literary venture, entirely different in scope and character from those discussed previously in this chapter, is revealing of an aspect of Kornberg's personality of which many are likely unaware. At a time when his three sons (who are separated in age by a mere three years) were very young, Kornberg not infrequently devoted time to inform and interest them in biology in general and the research that transpired in his own laboratories over the years in specific. Frustrated by attempts to educate them to the complexities of vitamins and enzymes, he playfully regaled them — and subsequently his grandchildren — with stories he concocted about the world of tiny organisms (little beasties). In his own words:

> I told my three sons stories about germs more than fifty years ago. Years later, on the several occasions when I took one and then another of my eight grandchildren on an extended lecture trip, their fathers urged that I tell them the Germ Stories. No longer able to concoct such tales, I instead tried to make instructive and entertaining rhymes and included in each poem a grandchild's name or that of a cousin.[51]

Shortly before he passed on in October 2007 Kornberg, at the urging of his sons, orchestrated the publication of a thoroughly delightful, charming and colorfully illustrated (by California artist Adam Alaniz) book entitled *Germ Stories,*[52] replete with many of Kornberg's poems. The book opens with a ditty called *The Germ Parade* that sets the tone for a dozen endearing "germ stories."

> Hurry, hurry to the parade
> Of the strangest creatures ever made
>
> No legs, no fins, no mouths, no eyes,
> Little beasties of the tiniest size
>
> Far too small for the eye to see -----
> "Just how small *is* this menagerie?"

Another, on poliovirus, begins:

> Long ago, summertime, with no more school
> There was lots of time in the swimming pool
> For Rog, Tom and Ken, brothers three,
> To play and frolic, trouble free
>
> But not so glad were Dad and Mum.
> The fear of polio made them glum
>
> They'd worry and fret at each sneeze of wheeze,
> Since with summer came this dread disease
>
> One year earlier, nineteen fifty-four,
> Polio struck at Steve, who lived next door.

Germ Stories, a collection of poems unmistakably intended for children, was reviewed in the *FASEB Journal* in 2008. The reviewer wrote:

> The work should be commended for trying to achieve the ambitious goal of making microbiology interesting and relevant to young children. For instance, the book has a poem that serves as a great conversation starter about polio — the threat it once was and the medical advances that have nearly eradicated the illness. Another great example is in the "Intestinal Menagerie", where Kornberg deftly explains the complex relationships among the bugs that make vitamins and the health of the people in whom they live.[53]

Sadly, Kornberg was not alive to see first-hand this lively publication, which he most certainly would have appreciated.

References

1. ECF interview with Roger Kornberg, May 2014.
2. Ibid.
3. Kornberg A. (1987) The Two Cultures Biology and Medicine. *Biochemistry* **26**: 6888–6891.
4. Roger Kornberg, personal communication.

5. Kornberg A. (1987) The Two Cultures Biology and Medicine. *Biochemistry* **26:** 6888–6891.
6. AK Interview with Sally Smith Hughes, Program in the History of the Biosciences and Biotechnology, Biochemistry at Stanford, Biotechnology at DNAX, Arthur Kornberg, 1997, p. 72.
7. Ibid., p. 73.
8. Kornberg A. (1988) DNA Replication, *Biochim Biophys Acta — Gene Structure and Expression* **951:** 235–239.
9. Kornberg A. (1991) *For the Love of Enzymes: The Odyssey of a Biochemist.* Harvard University Press, p. 1.
10. Kornberg A. (1996) Chemistry — the *lingua franca* of the Medical and Biological Sciences. *Chemistry and Biology* **3:** 3–5.
11. Kornberg A. (1997) Centenary of the Birth of Modern Biochemistry. *TIBS* **22:** 282–283.
12. Kornberg A. (1961) *Enzymatic Synthesis of DNA.* John Wiley & Sons, Inc.
13. Kornberg A. (1974) *DNA Synthesis.* W. H. Freeman and Company, San Francisco.
14. Kornberg A. (1991) *For the Love of Enzymes: The Odyssey of a Biochemist.* Harvard University Press, p. 303.
15. Ibid.
16. Kornberg A. (1980) *DNA Replication.* W. H. Freeman and Company, San Francisco.
17. Kornberg A, Baker TA. (1992) *DNA Replication 2nd Edition.* W. H. Freeman and Company, San Francisco.
18. ECF interview with Tania Baker, Nov. 2013.
19. Ibid.
20. Ibid.
21. Ibid.
22. Kornberg A. (1991) *For the Love of Enzymes: The Odyssey of a Biochemist.* Harvard University Press.
23. http://www.amazon.com/For-Love-Enzymes-Odyssey-Biochemist/dp/0674307763.
24. Joshua Lederberg, Introduction to AK Interview with Sally Smith Hughes, Program in the History of the Biosciences and Biotechnology, Biochemistry at Stanford, Biotechnology at DNAX, Arthur Kornberg, 1997, p. 2.
25. Kennedy E. (1989) Book Review — A Life in Biochemistry. *SCIENCE* **244:** 852–853.

26. ECF interview with Tom Kornberg, May 2014.
27. Kornberg A. (1991) Preface to *For the Love of Enzymes: The Odyssey of a Biochemist*. Harvard University Press.
28. Campbell PN. (1989) *Biochemical Education* **17:** 217.
29. Bondy P. (1990) For the Love of Enzymes. *Yale J Biol Med* **63:** 72–74.
30. Roger Kornberg, personal communication.
31. ECF Interview with A. Dale Kaiser, Aug. 2013.
32. ECF Interview with James Spudich, Oct. 2014.
33. ECF Interview with Robert Lehman, Oct. 2014.
34. Ibid.
35. Tauopathy. http://en.wikipedia.org/wiki/Tauopathy.
36. ECF Interview with Roger Kornberg, May 2014.
37. ECF Interview with Ken Kornberg, March 2014.
38. Kornberg A. (1991) Preface to *For the Love of Enzymes: The Odyssey of a Biochemist*. Harvard University Press.
39. ECF Interview with Paul Berg, Nov. 2013.
40. ECF Interview with Robert Lehman, Oct. 2014.
41. ECF Interview with Paul Berg, Nov. 2013.
42. Roger Kornberg, personal communication.
43. ECF interview with Tom Kornberg, May 2014.
44. ECF interview with Carolyn Kornberg, May 2014.
45. Ibid.
46. Kornberg A. (1995) *The Golden Helix: Inside Biotech Ventures*. University Science Books, Sausalito, California.
47. Kornberg A. (1995) Preface to *The Golden Helix: Inside Biotech Ventures*. University Science Books, Sausalito, California, p. x.
48. Schwartz RS. (1995) Book Review — The Golden Helix: Inside Biotech Venture. *New England J Med* **333:** 1292–1293.
49. Ibid.
50. Kornberg A. (1996) To the Editor, Review of Kornberg's The Golden Helix. *New England J Med* **334:** 994.
51. Kornberg A. (2007) Preface to *Germ Stories*. University Science Books, Sausalito, California.
52. Kornberg A. (2007) *Germ Stories*. University Science Books, Sausalito, California.
53. Mooneyhan C. (2008) A Parade of the Strangest Creatures Ever Made: A Review of *Germ Stories* by Arthur Kornberg. *FASEB J* **22:** 639.

CHAPTER FIFTEEN

In Support of Basic Science

Kornberg frequently deployed his pen to focus attention on government funding of biomedical research. In an editorial entitled "The Support of Science" published in the journal *SCIENCE* in June 1973, he drew attention to the financial attrition scientists were then facing.

> Funds for important research have been cut at a time when inflation and advanced technology increases: the support for the training of our best young scientists has been abruptly eliminated. This support for research and training cannot be finely regulated. When the flow of science support is turned down, the stream of progress dries up and cannot be restored for years.

"There are two compelling reasons why society must support basic science," he continued.

> One is substantial: the theoretical physics of yesterday is the nuclear defense of today; the obscure synthetic chemistry of yesterday is curing disease today. The other reason is cultural. The essence of our civilization is to explore and analyze the nature of man and his surroundings. As proclaimed in the Bible, in the Book of Proverbs: "Where there is no vision, the people perish." America's strength is not in mineral resources, in hydroelectric power, or in agriculture. It is not in the accumulation of a huge weapons arsenal either. America's strength is in the moral and intellectual resources of the people.[1]

It merits prefatory mention that young readers may not have witnessed the dire financial straits that threatened the viability of biomedical research support in the United States around the time of the Vietnam War and in other periods since. Indeed, in late 1967 Stanford University Vice-President and Provost Richard Lyman was prompted to write a lengthy memo to the leadership of all academic departments at Stanford addressing this distressing situation.

In the Fall of 1964 Kornberg (among unnamed others) was invited to the White House to meet with President Lyndon Johnson. The telegram he received from Donald M. Macarthur, Executive Director of Scientists and Engineers for Johnson and Humphrey, arrived a mere six days prior to the meeting. It states:

> A small group of nationally known scientists, engineers and physicians will express the President [to] our conviction that the continuance of enlightened policies of this and previous administration is essential for future welfare and progress of nation. Means for strengthening national program in science technology and health will be discussed.[2]

Johnson's diary for October 6, 1964 notes that he was scheduled to meet with the invited scientists, engineers and physicians at 11:55 am. Whether or not the meeting actually began at that precise time is a matter of conjecture, especially in light of a notation in the President's Daily Diary Worksheet for October 6, 1964 that at 11:53 am he was scheduled to meet or actually met with Jack Valenti and Dr. Donald Hornig, the latter being a Special Assistant to the President and Director of the Office of Science and Technology. It is not unreasonable to assume that any conversation between Valenti, Horning and Johnson lasted more than a mere two minutes. Johnson's Daily Diary Worksheet for October 6, 1964 also notes that at 12:25 pm the President "Returned to office — stopping in outer office to read the latest papers — *Washington Past* and *Evening Star*." One therefore assumes that little if anything of substance in the realm of science policy emerged from that meeting. However, it is conceivable that it was followed by one or more by other meetings at the White House that day. If so, no record of such was identified in Kornberg's Stanford archives.

As early as March 1968 Kornberg testified before the US Senate Subcommittee on Government Research, following which he wrote to the Director of the NIH, James Shannon.

> I am more convinced than ever that the American people must be better informed of the importance of basic research in the service of man and relieved of the anxieties that this investment will beget a monster. How to do this? _____ We cannot rely on scientists. They are with rare exceptions not suited, trained and/or interested. Scientists are, if anything, more self-centered and less concerned with broader social issues that affect their work than people in most any other craft or profession. You can start a fire down the hall but as long as the flames are not licking at his door, the serious scientist will ignore it and keep on working.[3]

The NIH budget for basic science was a pet peeve of Kornberg's, one that he attacked relentlessly. In a 1968 publication of a presentation he made at San Diego State University on "Genetic Chemistry and the Future of Medicine," he passionately offered one of many pleas he made in this regard. "Having been at the National Institutes of Health in Washington during the post-World War II period, I witnessed the birth and development of the federal programs that fueled the training and research that made Americans and the English language dominant in the world of science," he wrote.

> The miraculous success of the programs of the National Institutes of Health and the National Science Foundation, as impressive as anything our government has ever done, made me believe that they would be treasured and expanded. In this I was mistaken. For reasons of political ambition, envy and extraneous economic pressures, the support of basic research has been steadily eroded. The past two years have been especially bad. There are clear signs that industry is spending more on research, but these investments are in development rather than basic research. Any expectation that industry will replace the loss of federal funds is foolish. Industrial funds for the support of basic science are now less than five percent of the federal outlay and no one believes that this contribution will increase significantly.[4]

Such was the fervor of Kornberg's passion for increased financial support for basic biomedical research that during his lengthy and comprehensive 1997 interview with Sally Smith Hughes, he optimistically informed her that he was

> proposing a lobby of scientists of a magnitude that has not been envisioned, which will then have some voice in Washington. This is a lobby in the best sense of the term; it will inform Congress of the value of biomedical discoveries. That information will then be transferred to every community, where scientists will be available to meet with lay groups and explain why work on, let's say, the pathway of making building blocks of DNA, will lead to better drugs and better devices. One aim will be to dispel the illusion that basic research of any quantity and quality will be done in other than an academic setting. Ninety-five percent of the information that is going to be used in a fundamental way is not obtained in an industrial setting. Let me make it clear that all of the biotechnology that we have been talking about was done in laboratories that were built and supported by the NIH. It was not done in any commercial laboratory. All the ventures, people and ideas came from academic groups. That is still largely true.[5]

Later in the interview Kornberg informed Sally Smith Hughes:

> I've been talking for a year and a half about a truly major effort for federal support of biomedical research. There may be some movement. There would be more if I were twenty years younger and devoted my energies to it. We have to get leadership![6]

On another occasion Kornberg prepared a lengthy draft document co-signed by a number of prominent Stanford chemists and biochemists that addressed the establishment of a federally funded program to increase interaction between chemistry and biological research on protein-nucleic acids.

> "We believe that specific measures must be taken to stimulate and facilitate the efforts of chemists and chemistry departments to enlarge the scope of their research and research training in order to include more problems of current concern to the biologist and medical scientist,"

he wrote. "The suggestions offered in this report will, we believe, serve to promote accomplishment of this objective."

To the best of the author's knowledge this initiative stalled somewhere along the line, perhaps ultimately consigned to burial in the gargantuan collection of papers that now grace Kornberg's Stanford archives. But the mere fact of his once again taking time and energy to promote basic science amongst his many other commitments and responsibilities is testament to his deep concern about financial support of basic science.

Sometime in 1975 Kornberg was invited to join Senator Alan Cranston in a discussion of the major health issues facing California and the nation. In writing a letter of acceptance to participate in this initiative Kornberg put his finger on several critical issues that in his considered view had plagued the support of basic biomedical research for decades.

> "The difficulty with research support in our society, I have come to realize, is the failure to understand the nature and importance of basic research," he wrote in his letter. "This is true of the lay public and physicians, of legislators and political leaders. They do not realize the long time scale of basic research, and that its utility, its applicability, is not obvious. If it were it would be developmental rather than basic research. They do not realize the quantitative scale of basic research. Fragments of knowledge unwelcomed and unexploited are lost, as were Gregor Mendel's basic genetic discoveries. They also do not realize that the scientists who do basic research are the least articulate, least organized and least able to justify what they are doing, and this is in a society where selling is so important, where the medium is the message."[7]

Anxiety was seriously sharpened by President Richard Nixon's declaration of a *War on Cancer* in 1971, followed by the introduction in 1973 of a congressional bill introduced to amend the Public Health Service Act that had long provided support for research training by peer review through the NIH and the National Science Foundation (NSF) and replacing it with the *National Health Research Fellowship and Traineeship Act of 1973*. The essential purpose of this woeful proposition was to replace existing funding mechanisms for various specific training grants and or other sources of government spending that were modeled around different research areas and that had for years served trainees and their mentors well, with a new mode of funding.

A specific provision of the proposed Act stated:

> Any assistance provided any individual ———————— shall be provided on the condition that such individual will, in accordance with paragraph (2), engage in research or teaching for a twenty-four-month period for each academic year with respect to which he received such assistance. ———————————————— *The Secretary shall by regulation prescribe* (A) *the type of research and training which an individual may engage in to comply with such requirement, and* (B) *such other requirements respecting such research and training as he deems necessary.* (Author's italics.)

On March 22, 1973 Kornberg provided testimony at hearings on the Act. His scathing opening statement pulled no punches.

> Those of us who do research in medical science and train young people for research in medical science have witnessed in recent weeks the most calamitous decision a government of the United States could make for the future of medicine and the welfare of our country. Were there an intentional effort to undermine the health and economic welfare of this country for the coming generations, I could imagine nothing more devastating than to stop training our best young people to do research in basic medical science. Yet this is precisely what has been done and the consequences of the decision have not been foreseen.[8]

Mercifully, the bill never saw the light of day! A year later, with federal support for basic biomedical research still a hot button topic, Kornberg reminded the general public of President Richard Nixon's so-called *War on Cancer* with a perspective published in *The Denver Post* on July 7, 1974. Boldly entitled "How the 'Cancer War' is Wounding America," Kornberg's piece reiterated the sentiment that he "could imagine nothing more devastating than to stop training our best young people to do research in basic medical science," adding: "yet this is what President Nixon and the Secretary of Health, Education And Welfare, Casper

Weinberger, have done, despite earnest and repeated pleadings from the scientific community." A few days later Kornberg wrote a similar article for the *Los Angeles Times* with the heading "Cut in Basic Research Funds Called Tragic."

In the spring of 1975 Kornberg testified before a congressional committee about the state of basic science funding in the United States. The list of issues that he addressed included the following:

1. Lack of clarity in definitions of basic research and applied research
2. Concerns about federal spending in research and health care
3. Areas where federal spending in scientific matters should be reviewed
4. Concern about federal efforts to change the role and mission of the National Institutes of Health
5. Answering charges from some areas of government that important unapplied knowledge lies unused in laboratories and never reaches the sick
6. Concern about federal spending in research and health care

In the mid-late 1970s the issue of government "interference" in basic biomedical research surfaced again in a different context, this time in the controversy centering around the use of recombinant DNA (that had its origins at Stanford University in the late 1960s and early-mid 1970s), now one of the most powerful sets of research tools in the armamentarium of molecular biology.

In 1977 Kornberg corresponded with Congressman Paul N. McCloskey about issues that continued to flare up around government funding for basic research. He enjoyed a comfortable personal relationship with McCloskey, a Republican politician from California who served in the U.S. House of Representatives from 1967 to 1983. McCloskey ran on an anti-war platform for the Republican nomination for President in 1972 but was defeated by the incumbent Richard Nixon. "I realize that it is a time for belt tightening but not when the belt around the neck of research funding is already impeding circulation to its head," Kornberg wrote in his letter to McCloskey.[9]

In January 1978 Kornberg met with the late Senator Edward (Ted) Kennedy, then Chairman of the Subcommittee on Health and Scientific Research, a meeting that generated a stimulating exchange of views by correspondence. In a letter to Kornberg written on January 31, 1978, Kennedy encouragingly and astutely wrote:

> I have never questioned the intrinsic value of untargeted, investigator-initiated research, and I have always supported its generous funding at the National Science Foundation and at the National Institutes of Health. My commitment to investigator initiated research is based on my firm belief, which I have expressed repeatedly in hearings and speeches, that this variety of inquiry promises to yield real benefits for the health and quality of life of our people. As I recall, you argued the case of untargeted research on the same pragmatic grounds, so on this matter we start from the same premise.

"As you well know, however," Kennedy continued,

> it is much easier to espouse principles in the abstract than to implement them in practice. The promise of "basic" research is that it will yield knowledge which can be spun off in practical application. Clearly therefore, some portion of our research resources must be targeted on exploring the potential uses of new knowledge.[10]

One hopes that the halls of Congress in the 21st century hold others with the late senator's vision.

As recently as 1995 Kornberg was still publishing plaintive editorials and other documents of woe. "In the biomedical sciences we have become increasingly vulnerable to the prospect of severe cuts in federal support," he wrote in the journal *SCIENCE*.

> We must not let anyone be deluded into thinking that these cuts will be replaced to any significant extent by private and industrial sources of funding. In the period after World War II more than 90% of the support for the revolutionary advances made in the biomedical sciences came from the National Institutes of Health (NIH). No industrial

organization would have invested or ever will invest millions of dollars annually, for decades, in projects that have no direct relevance to marketable products.[11]

Earlier, in 1992, Kornberg published an address that he presented at a national symposium on *Today's Opportunities, Tomorrow's Health: The Future of Biomedical Research in America*. His address was subsequently published in the *FASEB* (Federation of American Societies for Experimental Biology) *Journal* (apparently the world's most cited biology journal) under the title "Basic Research: the Lifeline of Medicine."[12] Departing from his usual harsh rhetoric Kornberg singled out James Shannon, Director of the NIH from 1955–1968, "for his extraordinary vision and political skills that set NIH policy to support basic research," and he congratulated then NIH Director Bernadine Healy (who was present at the symposium) for establishing the Shannon Awards for Investigator-Initiated Research. He noted that during Shannon's watch the NIH budget grew 15-fold, from $81 million to $1.2 billion and pointed out that Shannon made it his business to see to it that the bulk of this funding was invested in basic research and training at the NIH and in universities around the world. "Consider for a moment the toll in suffering and waste of money had the bulk of it been spent compassionately on crusades to palliate these terrible diseases rather than investing in the basic understanding needed to prevent them," Kornberg wrote.[13]

Even research on DNA was considered a boon to human health. Kornberg pointed out that studies as basic as the enzymology of DNA "had no practical objective and seemed at the time to be irrelevant to producing products or procedures concerned with human welfare. Yet to my pleasant surprise such studies on the chemistry of DNA, its replication, repair, and rearrangements, have had a profound effect on our understanding of cellular processes and human diseases." He also pointedly noted that 10 years elapsed before Ernst Chain and Howard Florey discovered the therapeutic value of penicillin. Were Chain and Florey motivated by this practical goal? Not at all! "My final plea," Kornberg concluded, "is invest in science. It is as sound, practical and essential for our nation's health and industry as the investment we make in the rearing and education of our children."[14]

As discussed in the previous chapter on Kornberg's writing life, over a span of 40 years between 1964 and 2004 he penned a number of editorials, commentaries or articles that were either specifically directed at or referred to the fundamental importance of basic science in biomedical research. Many of these were prompted by talks and formal addresses he presented over the years. Others were spontaneously prompted by his unabated concern of this issue.

Kornberg fully understood that the community of scientists held as much if not more responsibility for the seesaw manner in which the vicissitudes of funding for basic biomedical research plague science. In his review of a book by Bernard Davis called *Storm Over Biology: Essays on Science, Sentiment, and Public Policy* published in *Nature* in 1986, Kornberg scathingly wrote:

> Scientists make poor politicians. They have been self-selected for their interest in things rather than people and they are generally neither literary nor articulate. They form societies to communicate with one another about their arcane findings, but with some exceptions they are distinguished from most other professional groups by their social isolation and political innocence. Were the responsibility of scientists limited to progress in such areas as chess or hieroglyphics their ineptitude in worldly matters might be tolerated with whimsical affection. But the affairs of government and business, our everyday lives and future depend crucially on science and its technological applications. Must we then rely on salesmen, actors and lawyers to make the difficult decisions that require informed judgment about scientific questions? What are we to do?[15]

Sally Smith Hughes broached the issue of funding for basic biomedical research during her 1997 interview with Kornberg, whose response was predictable, though laced with signs of weariness as his octogenarian years approached. He was then 79. "I think we should keep trying, even though I think it is almost hopeless," he told Hughes.

I have attempted to explain what DNA does, what it is and above that the importance of gaining basic knowledge to solve very practical problems in society, such as health and sanitation and environment. Does the public know that it is a good investment to support the whimsy of people who are curious about facts in nature? If you ask people, would they provide tax dollars for medical research, 75 or 80 percent say yes. Frame the question a little differently: major medical discoveries have been derived from the pursuit of curiosity of physicists, chemists, and biologists. Would you provide tax dollars for them to pursue their curiosity about facts in nature? They've not been asked that question. Maybe 10 or 20 percent would say yes.

My mission has been to cite chapter and verse as to how the drugs they are taking, the procedures they use — the MRI (magnetic resonance imaging), the X-rays, everything else they that they consider essential for their health — derives from such apparently irrelevant curiosity. And still the response the next day by them or their legislative representative is: "Hey, we've got AIDS; we've got cancer; we've got this or that disease. We can't afford to divert our limited budget, more stringent than ever, to support someone working on grasshoppers." Or as Bill Clinton purportedly said, "I'm not going to approve grants to work on stress in plants." Had he just thought about it one second more, he could have said, "Isn't it important that we cope with stress in plants — drought, disease, excessive moisture, lack of fertilizer?"

Unlike science where there is discernible progress, human understanding and social actions are not progressive. Still, scientists as responsible people, like others in society, should respond to opportunities to apply new knowledge to human welfare.[16]

References

1. Kornberg A. (1973) The Support of Science. *SCIENCE* 180: 909.
2. Telegram to Arthur Kornberg from Ronald M. Macarthur, Executive Director, Scientists and Engineers for Johnson and Humphrey.
3. Letter from Arthur Kornberg to James Shannon, March 11, 1968. Reproduced with permission from the Department of Special Collections, Stanford University Libraries.

4. Kornberg A. (1988) *Genetic Chemistry and the Future of Medicine.* San Diego State University Press, p. 15.
5. AK Interview with Sally Smith Hughes, Program in the History of the Biosciences and Biotechnology, Biochemistry at Stanford, Biotechnology at DNAX, Arthur Kornberg, 1997, pp. 54–55.
6. Ibid., p. 71.
7. Letter from Arthur Kornberg to Alan Cranston, Aug. 28, 1975. Reproduced with permission from the Department of Special Collections, Stanford University Libraries.
8. Testimony to US Congress, March 22, 1973.
9. Letter from Arthur Kornberg to Paul McCloskey, Jan. 7, 1976. Reproduced with permission from the Department of Special Collections, Stanford University Libraries.
10. Letter from Senator Edward Kennedy to Arthur Kornberg, Jan. 31, 1978. Reproduced with permission from the Department of Special Collections, Stanford University Libraries.
11. Kornberg A. (1995) Science in the Stationary Phase, *SCIENCE* 269: 1799.
12. Kornberg A. (1992) Basic Research: the Lifeline of Medicine. *FASEB J* 6: 3143–3145.
13. Ibid.
14. Ibid.
15. Kornberg A. (1986) What the Scientist Has to Say. *Nature* 324: 172.
16. AK Interview with Sally Smith Hughes, Program in the History of the Biosciences and Biotechnology, Biochemistry at Stanford, Biotechnology at DNAX, Arthur Kornberg, 1997, p. 70.

CHAPTER SIXTEEN

Commercial Ventures and the Founding of *DNAX*

Historically, the association of scientists with the private sector was largely, if not exclusively, limited to physicists and chemists. Only a small minority of the graduates of chemistry departments traditionally joined the academic sector. "Chemistry departments were tightly linked with industry," Kornberg told interviewer Sally Hughes. But until the decade of the '70s it was tacitly accepted that it was immoral for scientists in the biological research community to receive monetary reward for consultative activities. "In biology it was utterly unknown," Kornberg stated.[1]

Having found the few occasions on which he had consulted for pharmaceutical companies in the main "dissolutioning and disappointing," Kornberg had for many years shunned consultantships or any dealings with the pharmaceutical industry. "I had been distressed by several visits to pharmaceutical companies," he wrote.

> Typically, a team of five or ten pharmacologists and chemists was assigned a major disease category — cancer, arthritis, hypertension — and asked to develop patentable drugs to compete with those already in use. These scientists, once young and eager, had become gnomes grappling hopelessly with problems far beyond their reach.[2]

"It was not acceptable to degrade and prostitute yourself by engaging in activity that was done under such nonscientific, unproductive, intellectual circumstances. Occasionally there were genuine discoveries in industry, but they were relatively rare," Kornberg opined to Sally Smith Hughes.[3]

But by the late 1970s the recombinant DNA technology initiated in Paul Berg's laboratory greatly enhanced the Department of Biochemistry's reputation and it began to attract the attention of innovative pharmaceutical companies interested in profiting from its skills. The biochemistry faculty were aware of such a lucrative program in the Stanford Department of Chemistry and Berg, who had recently taken over the chairmanship of the biochemistry department, suggested that the department should follow suit. "We were very confident that we had significant molecular biological expertise to contribute to pharmaceutical and chemical companies," he stated. Kornberg was enthusiastically supportive and in turn wrote to Dale Kaiser stating: "I think we should consider this a top priority item and do it soon — if we do it at all. We're already a little late in getting into this sort of venture, and it may soon be too competitive to make it attractive."[4]

A maximum of three representatives per affiliate were allowed in any given year, with all travel expenses absorbed by the visitors. Each affiliate received a bound copy of the department's publications as well as preprints of papers in press for a given year. The annual affiliate membership fee was set at $12,000. When launched for the year 1980–1981, the program signed up 15 affiliates, including such notables as Bristol-Myers Company, Cetus Corporation, Genentech, Hoffman–LaRoche, Inc., Smith Kline & French Laboratories, and the Upjohn Company. Additionally, once a year a biochemistry faculty member was invited to visit an affiliate to present a research seminar and discuss subjects of mutual interest.

The biochemistry department hosted its first symposium for industrial affiliates, a two-day affair, in late March 1981. After much deliberation Kornberg decided not to invite students and postdoctoral fellows to facilitate keeping attendance at a manageable number. The symposium was an unqualified success. The second program was extended to three days and featured presentations by the senior faculty, on "Chromosomes and Genes," "Targeting of Proteins Into Eukaryotic Cells" and "Control of Gene Expression by Steroid Hormones."

"We would get maybe ten thousand dollars a year from each of a dozen affiliates," Kornberg told Sally Hughes.[5] Modest though they were, these funds provided a useful source of unrestricted revenue to the department, which was now in a position to supplement salaries of graduate students and postdoctoral fellows and to offer stipends to foreign students and post-graduates who otherwise were not eligible for scholarships. This new revenue stream also helped establish new faculty members, and "we could do other things that might have been difficult to achieve with government grants." In return, the selected companies were afforded access to new technologies and developments in the department. "It was viewed by both sides as an attractive kind of *quid pro quo* where we were offering something that was valuable, and getting compensated for it financially. ⎯⎯⎯⎯⎯ Finally, I think the program had an educational advantage."[6]

But in the late 1990s the program came under close scrutiny in the biochemistry department. "The program has dwindled and is now almost moribund," Kornberg told Sally Hughes in 1997.

> We were expending a lot of energy chasing a buck and it became increasingly more evident that in the 1990s we were not offering a unique product. Companies were dropping out and it was difficult to recruit new ones. So, somewhere along the line, we wondered whether it was a good business to be promoting.[7]

Eventually the Affiliates Program was dropped.

This was by no means the end, nor in fact the beginning of Kornberg's involvement with the burgeoning world of biotechnology, which in fact dates back to the late 1960s. "Until 1968, the idea of my being associated with an industrial enterprise was utterly unattractive; it offered so little scientific reward," Kornberg wrote in his book *The Golden Helix: Inside Biotechnology*.[8] But in 1968 Kornberg received and accepted an invitation from Alex Zafferoni, then a Mexican biotechnology entrepreneur, to join the scientific advisory board of a Palo Alto-based company called *ALZA*, dedicated to developing innovative drug-delivery systems.

Zaffaroni received a PhD degree in biochemistry from the University of Rochester in 1949. While at Rochester and in the years immediately following Zaffaroni cultivated an astute business sense and became interested in the innovative application of cutting edge pharmacology in the private sector. In due course he lent his skills to a small Mexican chemical company that specialized in steroid research. Under Zaffaroni's skilled management *Syntex Corporation* (as the company was called) became a major multinational Palo Alto-based pharmaceutical company that among other efforts pioneered the development of oral contraceptives. Eager to establish more effective and efficient ways to regulate the delivery of drugs and optimize their pharmacologic efficacy, Zaffaroni founded *ALZA* (a play on his name), a new company in Palo Alto, which soon became the world's leading drug-delivery technology company.[9]

Kornberg's decision to become involved in the private sector was entirely motivated by his relationship with Zaffaroni. "Our friendship was very firm and I admired his inventive genius and his entrepreneurial skills," he related.

> I had been aware of the latter from his exploits with *Syntex*. What Zaffaroni wanted to do I knew from my own medical training was very novel, that is, to deliver drugs in a programmed way to the place where they would be most effective, and to measure that delivery in very sophisticated chemical terms, exploring the physiologic as well as the chemical components. His most outstanding gift — there are several — is to inspire people to contribute and sustain their enthusiasm and expertise, and to create an environment where people help each other rather than compete with each other. .... On the other hand, he sees applications of science in a very novel and effective way, and in addition to this capacity to engage and to retain people in the enterprise he is also very shrewd in financing the enterprise. ....
> .... Everything he touches people assume will turn to gold, and so any venture he starts is oversubscribed.[10]

Alex Zaffaroni's business skills and acumen were recognized by his receipt of the National Medal of Technology from President Bill Clinton in 1995. In 2005 he received the Bower Award for Business Leadership from the Franklin Institute for his creation of new biochemical processes and drug delivery technologies. In the same year he was awarded the Gregory Pincus Award from the Worcester Foundation and in 2006 he

Alejandro Zaffaroni.

received the Biotechnology Heritage Award. Zaffaroni died in early March 2014 at the age of 91.

In the early days of the business relationships between the private sector and academia such affiliations were viewed with considerable scorn in the ivory tower, which considered such associations beneath the dignity of pure scientists, with no redeeming features other than generating personal financial income on the side. As already pointed out, Kornberg himself held this view at one point. But in mid-1981, a time when he was in active partnership with Alex Zaffaroni, he communicated a considerably more tempered attitude in a letter to Max Perutz, Director of the Laboratory of Molecular Biology (LMB) in Cambridge, UK.

> There has been so much written about the threat of industrial ventures in molecular biology to the academic enterprise. Most of it, to my mind, is nonsense. For example, there is a widespread notion that industrial operations require secrecy. This assumption is not valid. Whereas an academic competitor can use an idea or even appropriate an observation, this is most unlikely in industry. It has been Zaffaroni's belief and practice that secrecy makes even less sense than in academia; a successful product requires a commitment for many years and much money,

which, if protected by patents at appropriate stages, is not susceptible to being scooped. In any case, it will be interesting and possibly rewarding to see how these new ventures develop.[11]

Over the years Kornberg acquired increasing familiarity with the swiftly emerging biotechnology sector by serving on the scientific advisory boards of a multitude of companies, including *Regeneron*, *Metrigen*, *Xoma*, and *Galagen*. Surprisingly perhaps, he enjoyed these forays into the biotech world. "My associations with enterprises and people outside the traditional academic sphere have been personally gratifying," he wrote.[12]

In the early 1980s Channing Robertson, a faculty member in the Stanford Department of Chemical Engineering, and Alan Michaels, an experienced expert in membrane filtration technology, approached Kornberg, Berg and Charles Yanofsky in the Department of Biological Sciences with the notion of launching a Stanford industrial venture called *Engenics*, to be launched by a group of local venture capitalists.[13] When Zaffaroni learned that Stanford faculty of the ilk of Kornberg, Berg and Yanofsky were considering joining *Engenics* he invited them instead to join him in starting *DNAX*, a venture with the initial goal of developing genetically engineered antibodies.

"What precipitated the origin of *DNAX* was an invitation Alex had from the *Genetics Institute* in Boston that wanted him to direct it," Kornberg related. "As I recall, in October of 1980 when he broached this with me, I said to him: 'Why do you want to go to Boston and work with people who are really not that attractive. You could do it here.'" "Well, that's what I've been waiting to hear from you for some years," was Zaffaroni's delighted rejoinder.[14] With Zaffaroni's encouragement Kornberg invited Paul Berg and Charles Yanofsky to join as founders of *DNAX*. "We Stanford three — Paul, Charley, and I — brought our experience and outlook in science to complement Alex's entrepreneurial talents and vision," Kornberg wrote.

> Paul was a discoverer of recombinant DNA, a leader in enzymology and molecular biology of protein synthesis, and at the forefront of the genetic engineering of vectors powered by tumor-virus genes. Charley,

Charles Yanofsky.

equally proficient in biochemistry and genetics, had promptly adopted and advanced recombinant DNA techniques to clarify still further the mechanisms and regulation of gene expression. I had discovered and characterized many of the enzymes of DNA replication that later became essential reagents in gene splicing and other DNA manipulations.[15]

The scientific talent of this trio was further complemented by the recruitment from Japan of Kenichi Arai and his wife Naoko, former postdoctoral fellows in Kornberg's Stanford laboratory.[16]

Kornberg, Berg, Yanofsky and Zaffaroni occasionally met for lunchtime chats to evaluate advances in the enzymology of DNA replication, DNA repair and genetic recombination. Zaffaroni was a keen listener, especially to the discussions of the advances on recombinant DNA technology made in Paul Berg's and Stanley Cohen's Stanford laboratories. As Kornberg put it:

> Despite his muted way of articulating his visions and plans, Alex manages to excite scientists and businessmen alike and to sustain their devotion. There was complete unanimity among the group on certain irrefutable guiding principles. We would select the best young scientists

we could find, offer them the resources to work diligently and creatively, and trust that, in time, their associations with one another would lead to common interests, and the identification of common goals.

Initially we would not compete with those ventures already well started in the cloning of a number of known factors ──────────────. We decided to exploit biotechnology by going after antibodies, which could be obtained in quantity by the recently developed monoclonal-antibody technique. We would hone and redesign them in order to solve medical problems and, perhaps, even to serve as some industrial purposes. With advice from medical colleagues and with our experience in molecular biology and protein chemistry, we would select suitable medical targets for antibodies and direct the engineering to make them.[17]

A requirement that the company be close to Stanford was obligatory. Zaffaroni identified a building that Alza owned in the Stanford Industrial Park, a mere five-minute drive from the medical school. Kornberg's initial dismay on viewing the "dilapidated and cavernous spaces was allayed by Zaffaroni's assurance that a clever architect and a million dollars would convert the 10,000 square feet into attractive laboratories and offices."

"Knowledge for its own sake is not the ideal recipe for a business, so it has to be leavened — or unleavened — with the practical reality of getting a product — and a profitable marketable product," Kornberg told interviewer Sally Hughes.

And that's where Alex had a marvelous touch. He would not accept something that was profitable through gimmickry or outright copying of something that is profitable. So in the annals of technology it had to be a genuine application of basic technology. In the case of *DNAX* it had to be an important protein that meets all these standards and does something novel.[18]

The choice of an architect was a no-brainer and Arthur's son Ken Kornberg eagerly agreed to design the new entity. Rather than the conventional, straight, dim hallways, passages were angular and brightened by clerestory windows and the generous use of splashes of color. Scientific staffing of a new biotech venture in the early 1980s was not trivial. The uncertain shape and future of a new enterprise and the stigma attached to an industrial position (when compared to an academic appointment) were daunting obstacles. But some scientists saw in *DNAX* a favorable setting with access to all the requirements to make their mark as scientists. Financial inducements of equity in the company and higher salaries than those offered in academia were secondary considerations.

The first wave of recruits included Kenichi and Naoko Arai, a husband and wife team that had previously worked in Kornberg's Stanford laboratory as postdoctoral fellows. Kenichi Arai had acquired an MD degree from the University of Tokyo in 1967 and subsequently undertook graduate research with Yoshito Kaziro, a world-class scientist and one of the foremost biochemists in Japan, who Kornberg first met at the symposium in Madrid to celebrate Severo Ochoa's 70th birthday in 1975 (see Chapter 2). Arai's PhD thesis and subsequent work in a junior faculty position contributed to major discoveries of the mechanism of protein synthesis and the transmission of hormonal and cytokine signals by GTP-binding proteins.[19]

Naoko Arai in turn, who attended a private girl's school in Tokyo associated with the Japanese Women's University, obtained her MD

Ken-Ichi Arai.

Naoko Arai.

degree from Yokohama University in 1971, where she rekindled her relationship with Kenichi, who she knew as a marriage-related cousin. Like Kenichi, Naoko subsequently obtained her PhD under Kaziro. Naoko had earlier deferred a marriage proposal from Kenichi in order to complete medical school. But the two married when they were both in Yashito Kaziro's laboratory. Like her husband, Naoko Arai pursued further work in Kaziro's laboratory as a postdoctoral fellow and subsequently as a junior faculty member.[20]

"They came to my laboratory directly from the airport, Kenichi with a clipboard listing multiple chores and detailed research plans," Kornberg wrote. An ardent sports fan, Kornberg commented that "they reminded me, in an offbeat way, of Bill Walsh, the celebrated coach of the San Francisco 49ers, who listed on his clipboard the first twenty-five plays he would call in the game that day."[21]

Those with even a modicum of understanding of Japanese academic culture will well understand that it took considerable courage for Kenichi to confront Professor Yashito Kaziro, his Japanese mentor, who wanted and needed Arai in Japan. Kaziro was distressed and disappointed to lose his star student and it took several years before Kenichi, vindicated by research success, returned to Japan and restored close relations with his mentor.[22]

Kornberg described Kenichi Arai as a dynamo.

> His energy and intellect are expressed in a remarkable capacity to master the basics and details of a wide gamut of biological disciplines and in an ability to collect, galvanize and propel the research efforts of a team of pupils and peers. His scholarly output during his three post-doctoral years in the Kornberg laboratory — 10 papers in a single issue of the *Journal of Biological Chemistry* — attracted high level attention in academic biochemical circles and both he and Naoko had been plied with job offers in academic institutions in the United States, including a full professorship at Columbia University, before returning to Japan.[23]

Upon returning to the University of Tokyo following her postdoctoral training with Kornberg Naoko was unable to find an academic position and worked without status or salary, while for Kenichi, the resources and independence as an assistant professor were woefully meager. Kornberg noted that

> government support of facilities, training, and basic research, even at the most elite of the Japanese universities, had been, and was likely to remain, abysmal. ──────── Kenichi and Naoko were thus ready to accept my invitation to come to *DNAX*.[24]

At a November 1983 meeting of the faculty in the Stanford Department of Biochemistry it was decided to award Kenichi Arai a consulting professorship. When questioned about this apparent deviation from Kornberg's long-held policy of not providing faculty appointments to individuals other than full-time tenure-accruing individuals, Kornberg characterized the appointment of Arai as a consulting professor as a trivial appointment that carried no obligation by the department. "On the other hand, you don't hand out these appointments without serious concern about the desirability and the credibility of the person," he stated.[25]

Others recruited to *DNAX* had equally stellar resumes. Bob Coffman was an immunologist who did postdoctoral work in Irving Weisman's Stanford laboratory; Frank Lee completed his PhD training under Charles Yanofsky in the Stanford Department of Biology Department and turned down a junior faculty appointment at Columbia University to join *DNAX*; Kevin Moore, a Princeton graduate, obtained his PhD at CalTech and pursued postdoctoral training with Lee Hood at Caltech; Gerard Zurawski an Australian, did postdoctoral work in Charles Yanofsky laboratory at Stanford, and was encouraged by Lee Hood to join *DNAX* instead; Tim Mossman, relinquished a faculty position at the University of Alberta; and Donna Renick, who despite many impediments, notably the strong misgivings of her beloved and respected thesis advisor Eliezer Benjamini at the University of California, Davis but whose work on antiidiotype antibodies was attuned to *DNAX*'s project and was in turn attracted to *DNAX* by its goals and philosophy.[26]

Operating from his ALZA office, Zaffaroni took on the business affairs and with advice from Kornberg, Berg and Yanofsky "named the new company, recruited scientists and staff, and selected initial projects."[27] In addition to Berg, Yanofsky and Kornberg the advisory board included a covey of scientific luminaries, including Leroy (Lee) Hood, Michael Hunkapiller, Roger and Tom Kornberg, Ronald Levy, Irving Weismann, Harden McConnell and others. Weisman, a faculty member in the Stanford Department of Pathology and presently the director of the Stanford Stem Cell Center, had served as a member of Schering–Plough's task force that recommended its foray into immunology and was a leading figure in the basic mechanisms of immune modulation.[28]

DNAX was launched with $4 million in cash assets. Kornberg pointed out that the average cost per scientist in a biotech venture in 1981 was, by rule of thumb, about $100,000 per year. But with remodeling of the laboratories, the purchase of equipment and other expense, the remaining funds were only sufficient to sustain a staff of 24 people for one more year. "Alex operates in a generous style, but even skimping to reduce the 'burn rate' could extend the life of *DNAX* for only a matter of months," Kornberg related. "We urgently needed more money, and 1981 was an especially lean year for biotech investments.[29] Kornberg and Zafferoni travelled to Japan, Paris and London on a fund raising mission (which raised the hackles of the reviewer of Kornberg's book *The Golden Helix* discussed in the Chapter 14), and held meetings with potential investors in the California Bay Area and other American pharmaceutical companies. "Alex continued the search for funding through the end of 1981," Kornberg wrote.[30]

Numerous entreaties that engaged multiple established companies yielded no overt interest in investing in *DNAX*. Kornberg was clearly disappointed. When he complained to Zaffaroni that, in nearly 30 years of applying for grants, he had never been turned down, Alex commented: "Get used to it. I was turned down more than 40 times seeking help for *ALZA*." "Thus did my internship in financing commence," Kornberg humorously retorted. "I was learning that even powerful and exciting scientific ideas and approaches may lack the strength to open an investor's wallet."[31]

> But I cannot recall that Alex expressed any anxiety, even when the company's money would soon run out. He was confident that something would happen, and it did. Our savior was Schering–Plough. What brought them to *DNAX* and ultimately to the marriage altar were the people on each side — Alex and his colleagues on ours, Bob Luciano (who served as Chief Executive Officer of Schering–Plough Corporation from 1982 to 1996, as President from 1980 to 1986, and as Chairman of the Board from 1984 to October 1998) and his associates on theirs. Congruence of ideas and plans was essential but hardly sufficient. What mattered was the mutual trust built upon past performances.[32]

Within 18 months *DNAX* was acquired by Schering–Plough.

"During the period that the acquisition of *DNAX* by Schering–Plough was being negotiated the major research emphasis at *DNAX* was changed to fit with some of what Schering–Plough desired," Allan Waitz, a former president and CEO of *DNAX* stated.[33]

> "Schering was then dissatisfied with the small research efforts ongoing in immunology in their own laboratories in New Jersey, as well with the early efforts of a research laboratory they maintained in Lyon,

France and they convened an outside group of experts to recommend appropriate directions," Waitz related. "At that time, some research groups at Schering–Plough were beginning to grow specific mouse T cells in culture and the recommendation of the consulting group was that Schering focus on this and attempt to understand signals between various cells of the immune system; efforts that might lead to product opportunities and/or new tools related to three of Schering's primary areas of commercial product emphasis — allergy, inflammation and infectious disease."[34]

"The powers that be at Schering–Plough decided that collaboration with *DNAX* might offer a potential route to pursue these goals, and discussions were initiated with *DNAX* scientists to determine how they might put together a research effort that would satisfy the interests of Schering–Plough," Waitz continued. "At the time of the acquisition by Schering–Plough a research program to pursue an understanding of several lymphokines, their receptors, and how they function was put in place. It was considered that this might lead to molecules of intrinsic interest and/or to tools and understanding that might enable drug screening efforts in New Jersey to establish novel screens for pharmacologically valuable molecules."[35]

"Some of the research efforts at *DNAX* included the goal of cloning genes that encoded receptors specifically for cytokines and ultimately to understand the pathway by which signals are transmitted from receptors to the nucleus," Paul Berg told Sally Smith Hughes during an interview conducted in 1997.[36] "*DNAX* had an edge, a very special skill, in isolating and focusing on rare cells of the immune system," Berg stated. "If one could begin to enrich for certain cell types one might gain access to specific proteins in the immune system that would otherwise be extremely difficult to identify. And that was *DNAX*'s expertise in the late 1990s."[37]

In the opinion of Paul Berg the basic science infrastructure at *DNAX* at that time rivaled that of the famous Basel Institute of Immunology in Switzerland.

> Everybody was assured that they had a certain fraction of time, 20, 30, 40 percent of the time when they could do whatever they wanted.

However, their other research was to relate to the principle theme of *DNAX*. But the principle theme was very broad. It was molecular immunology and subsequently became molecular immunology and cell growth control. Within those boundaries, people could do almost anything.[38]

"We had excellent academic people in the company," Berg related.

A lot of young folks made their scientific reputations there. Alex Zaffaroni operated on the philosophy that it was the company's responsibility to encourage scientists to publish their research and to present their findings at scientific meetings. Essentially, if *DNAX* lawyers were required to file a patent on a Saturday night so that one of the scientists could make a presentation at a scientific meeting the following Monday — that's the way it was done.[39]

In the final analysis *DNAX* never fulfilled the hopes and dreams that Kornberg and Zaffaroni had for it. In November 2009 Schering–Plough merged with Merck & Co. and through a reverse merger Merck became a subsidiary of Schering–Plough, which renamed itself Merck. "*DNAX* no longer exists as such," Allan Waitz related in April 2015. "Kenichi moved with much of his group back to Japan and the program at *DNAX* became more of a drug screening effort."[40]

References

1. AK Interview with Sally Smith Hughes, Program in the History of the Biosciences and Biotechnology, Biochemistry at Stanford, Biotechnology at DNAX, Arthur Kornberg, 1997, p. 49.
2. Kornberg A. (2002) *The Golden Helix: Inside Biotech Ventures*. University Science Books, p. 26.
3. AK Interview with Sally Smith Hughes, Program in the History of the Biosciences and Biotechnology, Biochemistry at Stanford, Biotechnology at DNAX, Arthur Kornberg, 1997, p. 50.
4. Memo from Arthur Kornberg to Dale Kaiser, Sept. 10, 1979. Reproduced with permission from the Department of Special Collections, Stanford University Libraries.

5. AK Interview with Sally Smith Hughes, Program in the History of the Biosciences and Biotechnology, Biochemistry at Stanford, Biotechnology at DNAX, Arthur Kornberg, 1997, p. 56.
6. Ibid.
7. Ibid., p. 57.
8. Kornberg A. (2002) *The Golden Helix: Inside Biotech Ventures*. University Science Books, p. 26.
9. Ibid., pp. 59–97.
10. AK Interview with Sally Smith Hughes, Program in the History of the Biosciences and Biotechnology, Biochemistry at Stanford, Biotechnology at DNAX, Arthur Kornberg, 1997, p. 121.
11. Letter from Arthur Kornberg to Max Perutz, June 1, 1981.
12. Kornberg A. (2002) *The Golden Helix: Inside Biotech Ventures*. University Science Books, p. 27.
13. Ibid, p. 28–29.
14. Ibid, p. 29.
15. Ibid.
16. Ibid., p. 32.
17. Ibid., p. 30.
18. AK Interview with Sally Smith Hughes, Program in the History of the Biosciences and Biotechnology, Biochemistry at Stanford, Biotechnology at DNAX, Arthur Kornberg, 1997, p. 124.
19. Kornberg A. (2002) *The Golden Helix: Inside Biotech Ventures*, University Science Books, pp. 32–33.
20. Ibid., p. 34.
21. Ibid., p. 32.
22. Ibid., p. 36.
23. Ibid., p. 35.
24. Ibid., pp. 35, 36.
25. AK Interview with Sally Smith Hughes, Program in the History of the Biosciences and Biotechnology, Biochemistry at Stanford, Biotechnology at DNAX, Arthur Kornberg, 1997, p. 136.
26. Kornberg A. (2002) *The Golden Helix: Inside Biotech Ventures*. University Science Books, pp. 36–44.
27. Ibid., p. 29.
28. Ibid., p. 47.
29. Ibid., p. 54.

30. Ibid., pp. 56–57.
31. Ibid., p. 56.
32. Ibid., p. 57.
33. Allan Waitz, personal communication, April 2015.
34. Ibid.
35. Ibid.
36. Paul Berg Interview with Sally Smith Hughes, Program in the History of the Biosciences and Biotechnology, Biochemistry at Stanford, Biotechnology at DNAX, Arthur Kornberg, 1997, p. 166.
37. Ibid.
38. Ibid.
39. Friedberg EC. (2014) *A Biography of Paul Berg — The Recombinant DNA Controversy Revisited.* World Scientific Publishing Co., Singapore, p. 342.
40. Allan Waitz, personal communication, April 2015.

CHAPTER SEVENTEEN

A Life Well Lived

Arthur Kornberg was at the lab bench probing the secrets of polyphosphate until very shortly before his death on October 26, 2007. "One Friday afternoon his secretary at Stanford called me and told me that he was not doing too well," his wife Carolyn related.

> So I picked him up and took him home and put him to bed. A few days later he fell in the bathroom and I called 911. I contacted the boys right away. I think Roger was travelling. Arthur was checked into the hospital — and never came home. He was basically unconscious for the rest of the week and died the following Thursday. His heart and lungs were failing. He was 89 and just worn out! We took him off life support. He didn't want to be buried next to Charlene so he was cremated.[1]

Kornberg's death sparked a massive outpouring of tributes, too many to mention here. The Stanford Department of Biochemistry remembered him with a four-hour "teach-in" for which students and faculty selected favorite papers from Arthur's list of 463 publications and gave 5 to 10 minute summaries.[2] Paul Berg ended his tribute published in the *Proceedings of the American Philosophical Society* as follows: "Arthur Kornberg was the most influential person in my scientific career, and, most assuredly, many of the values I hold and actions I took over the years reflected what I had absorbed during our association of nearly fifty-five years"[3]

Arthur Kornberg in his Stanford Office.

In the words of his son Tom:

Arthur was passionate, resolute and totally committed to basic research and to the primacy of reason. He was strong and effective — with Senators and Congressman, benefactors, students, postdocs, family, etc. He had a dominating and inveterate optimism, and a positive outlook that was infectious and inspiring. He expected everyone to succeed and could invest himself in others in a way that compelled them to excel.[4]

As eloquent as any of the huge outpouring of comments about Arthur Kornberg was one published by his long-term and close friends and scientific colleagues, Paul Berg and Bob Lehman. "Both of us knew Arthur for more than 50 years, from the time we joined his laboratory at Washington University as postdoctoral fellows" the pair wrote.

"Our relationships with him went beyond that of student and mentor. We were embraced as members of his family and shared many special occasions and achievements that they celebrated. Arthur's style of doing science, his passion for experimentation rather than theory, and excitement about discovery inspired us. We remember the late-night calls inquiring how our experiments had fared. He was a serious and superb teacher and a generous and compassionate leader. The success of the faculties he assembled attests to his gift of forsaking the limelight and encouraging his colleagues to flourish on their own. Above all, Arthur was devoted to his students and colleagues and fiercely loyal to his family and friends. Perhaps Arthur's greatest legacy, and certainly the one of which he was most proud, was his extraordinary family of three sons and eight grandchildren. We will miss him greatly."[5]

Another published tribute, this on the occasion of Kornberg's 70th birthday in 1998, acknowledged his "greatest love," his first wife Sylvy Kornberg.

> We wish to recall the memory of Sylvy Kornberg, a devoted wife and mother, a gentle and gracious woman, and a fine biochemist. Those who were privileged to know her will forever be enriched by the memory of her friendship, her kindness, and her unfailing good humor.[6]

Kornberg completed his autobiography *For the Love of Enzymes: The Odyssey of a Biochemist* with a chapter entitled *Reflections of My Life in Science*. In this chapter he briefly revisited two issues about which he had especially passionate opinions — the import to him he obviously wished to imprint on his readers. As already mentioned, one issue was labeled *The Virus of Anti-Semitism;*[7] the other was entitled *Basic Research, the Lifeline of Medicine*.[8] Basic research was the essence of who Arthur Kornberg was — and what he stood for.

As pointed out by Kornberg's eldest son Roger, a keen historian of contemporary biochemistry and molecular biology,

> much as Francis Crick was a seminal leader in the field of molecular biology, so Arthur was the outstanding figure in nucleic acid biochemistry in his time. It was a mantle he inherited from Otto Heinrich Warburg, Otto Meyerhof, and the Coris. He led by force of example, and by a laser-like focus on the most central elements of the enzymology of the nucleic acids, which culminated very directly in the emergence of recombinant DNA technology and genetic engineering. The emergence of these crucial tools and disciplines is unmistakably a consequence of

Arthur Kornberg — the sunset years.

Kornberg's leadership, not only for his contributions to his development of the enzymology, but also for his creation of a school in which the ideas developed and the work was done.[9]

"Paul Berg is said to have referred to Arthur Kornberg as a 'legend in his own time.'" Roger Kornberg continued.

> He was that and more. He led not only by force of example but also by force of his personality. His insistence on excellence not only in his own work but in that of others permeated the entire field. Many of the leading figures looked to him for approval, seeking to live up to the standards that he not only set for them but impressed on them directly.[10]

Roger relates that Daniel (Dan) Koshland, the renowned American biochemist who proposed the induced fit model for enzyme catalysis and is credited for spearheading the reorganization of the biological sciences at the University of California at Berkeley,

> used to complain that Arthur Kornberg was critical of him because Arthur was always telling him that he could do better; until a time

when Koshland became the recipient of awards for which he discovered Arthur was responsible, and realized that Arthur Kornberg was the rare individual who would criticize him to his face — but laud him behind his back.[11]

Outside the laboratory Arthur Kornberg was also revered by his family. In the words of their youngest son Ken:

Arthur and Sylvy created a simple environment of trust and protection for their family that was based on honesty and affection. This extended outside our home to the many in his professional life. It was an umbrella that I still feel everyday, 29 years after Sylvy's death and 8 years after Arthur's.[12]

References

1. ECF interview with Carolyn Kornberg, May 2014.
2. Baldwin R. (2008) Recollections of Arthur Kornberg. *Protein Science* **17**: 385–388.
3. Berg P. (2009) Arthur Kornberg. *Proc Amer Phil Soc* **53**: 467–475.
4. Tom Kornberg, personal communication.
5. Berg P, Lehman IR. Memorial Resolution. (2007) Arthur Kornberg (1918–2007).
6. Berg P, Lehman R. (1988) Nucleic Acids Res **16(14A)**: 6263–6264.
7. Kornberg A. (1991) For the Love of Enzymes: The Odyssey of a Biochemist. Harvard University Press, pp. 310–313.
8. Ibid., pp. 314–318.
9. Roger Kornberg, personal communication.
10. Ibid.
11. Ibid.
12. Ken Kornberg, personal communication.

Appendices

Career Highlights

1937 Receives a bachelor degree from City College of New York and enters the University of Rochester School of Medicine

1941 Receives the MD degree from the University of Rochester and serves an internship in Internal Medicine at Strong Memorial Hospital, New York

1942 Enters the US Public Health Service and is assigned to the US Coast Guard

1942 Transfers to the Nutrition Section of the National Institutes of Health (NIH)

1946 Postdoctoral training with Severo Ochoa at New York University

1947 Postdoctoral training with Carl and Gerty Cori at Washington University, St. Louis

1947 Returns to the NIH and is named Chief, Enzyme and Metabolic Section, National Institute of Arthritis and Metabolic Diseases

1953 Professor and Chair, Department of Microbiology, Washington University, St. Louis

1956 Discovers DNA polymerase and achieves first laboratory synthesis of DNA

1959 Professor and Chair, Department of Biochemistry, Stanford University

1959 Shares the Nobel Prize in Physiology or Medicine with Severo Ochoa

1967 Achieves first replication of viable viral DNA

1969	Relinquishes chairmanship of the Stanford Department of Biochemistry
1980	Co-founder, *DNAX* Research Institute of Molecular and Cellular Biology, Palo Alto, California
1988	Assumes Emeritus status at Stanford University
1988	Shifts research focus to the study of inorganic polyphosphate
1999	Dedication of the Arthur Kornberg Medical Research Building, University of Rochester

Major Awards

1951	Paul–Lewis Award in Enzyme Chemistry
1959	Nobel Prize in Physiology or Medicine
1970	Elected to the American Philosophical Society
1970	Elected a Foreign Fellow of the Royal Society
1979	National Medal of Science

Honorary Degrees Received

City College of New York

Washington University

University of Rochester

Yeshiva University

University of Pennsylvania

Princeton University

Colby College

University of Notre Dame

University of Barcelona

University of Paris

Medical College of Wisconsin

University of Miami

Authored Books

Enzymatic Synthesis of DNA	John Wiley & Sons, (1961)
DNA Synthesis	W.H. Freeman and Co., San Francisco (1974)
DNA Replication	W.H. Freeman and Co., San Francisco (1980)
For the Love of Enzymes: The Odyssey of a Biochemist	Harvard University Press (1989)
DNA Replication, 2nd edition (with Tania A. Baker)	W.H. Freeman and Co., San Francisco (1992)
The Golden Helix: Inside Biotech Ventures	University Science Books (1995)
Germ Stories (with Adam Alaniz)	University Science Books (1995)

Essay, Editorials and Other Writings

"The Recent Revolution in Biology"	Essay based on 1968 Founders Day Address
"Basic Motives of a Professional"	*profiles.nlm.nih.gov/ps/access/WHBBHN.pdf*
"Life"	Based on an address delivered at Commencement Exercises of the University of Utah College of Medicine, June (1969)
"The Support of Science"	Editorial, *SCIENCE* **180**: 909 (1973)
"NIH, Alma Mater"	*The Pharos of Alpha Omegas Alpha* **38**: 98–101 (1975)
"National Institutes of Science Alma Mater"	*Science* **189**: 599 (1975)
"Nutrition and Science"	Food, Man and Society, Plenum Press, NY, 1–11 (1976)
"Research, the Lifeline of Medicine"	*NEJM* **294**: 1212–1216, May 27 (1976)
"For the Love of Enzymes"	Reflections on Biochemistry, Pergamon Press, NY 243–251 (1976)

"Severo Ochoa, Trabajos Reunidos de Severo Ochoa, 1928–1975"	*TIBS* **1:** 267–268 (1976)
"Future Trends — Why Work on the Enzymology of DNA Replication?"	*TIBS* **2:** N56–N57 (1977)
"Does a Doctor Need to Know Biochemistry?"	*TIBS* **3:** N73–N74 (1978)
"Biochemistry Evolving"	*Canadian J Biochem* **58:** 93–96 (1980)
"What the Scientist Has to Say"	*Nature* **324:** 172 (1986)
"The Two Cultures: Chemistry and Biology"	*Biochemistry* **26:** 6888–6891 (1987)
"Marketing of DNA"	In: Vision and Values for Pharmaceutical Innovation (C. Mitchell, ed.) 20th Anniversary Symposium, ALZA Corporation, 105–121 (June 1988)
"Understanding Life as Chemistry"	*Clinical Chemistry* **37:** 1895–1899 (1991)
"Understanding Life and Nutrition As Chemistry"	New Technologies and the Future of Food and Nutrition, J. E. Gaull, (ed.) John Wiley & Sons 59–69 (1991)
"A Dialogue With Linus Pauling and Daisaku Ikeda"	Foreword, *A Lifelong Quest For Peace*, Jones and Bartlett, Boston (1992)
"Science is Great, But Scientists Are Still People"	Editorial, *Science* **257:** 859 (1992)
"Basic Research: The Lifeline of Medicine"	*FASEB J News and Features* **6:** 3143–3145 (1992)
"Enzymology, Then and Now"	Henk Janz, Onderzoeker Pur Sang, Vakgroep Fysiologische Chemie, Utrecht, 139–142, (1992)
"Severo Ochoa (1905–1993)"	*Nature* **366:** 408 (1993)
"Understanding Life as Chemistry"	*Intl J Quantum Chem* **53:** 125–130 (1995)
"Science in the Stationary Phase"	*Science* **269:** 1799 (1995)

"Fifty Years Ago: The State of Biochemistry. Milestones in Biological Research"	*FASEB J* **9:** 1497–1498 (1995)
"Of Serendipity and Science"	Stanford Magazine: Summer (1995)
"Basic Research, The Lifeline of Medicine"	Essay for the Electronic Nobel Medicine. Nov. (1996)
"Support for Basic Biomedical" "Research: How Scientific Breakthroughs Occur"	*The Future of Biomedical Research* C. E. Barfield and B. L. R. Smith, eds., Washington, DC: American Enterprise Institute and the Brookings Institution (1997)
"Centenary of the Birth of Modern Biochemistry"	*TIBS* **22:** 282–283 (1997)
"Science and Medicine at the Millenium"	*Biochem Molec Med* **61:** 121–126 (1997)
"Severo Ochoa"	*Proc of the Amer Philosoph Soc* **141:** 479–491 (1997)
"The NIH Did It"	*Science* **278:** 1863 (1997)
"Centenary of the Birth of Modern Biochemistry"	*FASEB J* **11:** 1209–1214 (1997)
"Science and Medicine at the Millenium"	*Brazilian J Med Biolog Res* **30:** 1379–1386 (1997)
"Biotechnology in Science. Medicine and Industry"	*BioTecnoligia* **4:** 13–19 (1999)
"Future is Invented, Not Predicted"	*From the CAUT Bulletin*, Dec. 2000 (www.caut.ca)
"Tribute to Professor Igor S. Kulaev"	*Biochemistry* (Moscow) **65:** 279. (Translated from *Biokhimiya* **65:** No. 3 (2000)
"Remembering Our Teachers: in a series commissioned to celebrate the centenary of the JBC in 2005; JBC Centennial 1905–2005, 100 Years of Biochemistry and Molecular Biology"	*J Biol Chem* **276:** 3–11(2001)

"Public Funding of Understanding Nature"	*IUBMB Life* **51:** 71–72 (2001)
"How I Became a Biochemist"	*IUBMB Life* **53:** 185–186 (2002)
"Whither Biotechnology in Japan?"	*Harvard Asia Pacific* **6:** 6–9 (2002)
"Biochemistry Matters"	*Nature Structural & Molecular Biology* **11:** 493 (2004)
"Osamu Hayaishi: Pioneer First of The Oxygenases, Then the Molecular Basis of Sleep and Throughout a Great Statesman"	*IUBMB Life* **5:** 253 May–June (2006)

Index

Note: Page numbers in *italics* indicate photographs; those with "n" indicate footnotes.

A
Aab Institute of Biomedical Sciences, 125
Abraham Lincoln High School, 4, 5, 60
Abrams, Herbert, 116, 117
"Abundant Microbial Inorganic Polyphosphate, PolyP Kinase Are Underappreciated" (Kornberg), 227
Acetylcholine, 36
Aconitase, 26, 27, 31
Adenine, 73, 81, 90
Adenine ribose phosphate (ARP), 43, 73
Adenosine diphosphate (ADP), 25, 39, 81, 82, 121, 211
Adenosine monophosphate (AMP), 40, 41
Adenosine triphosphate (ATP), 20, 22–26, 38–40, 44, 45, 73, 81, 82, 88, 211, 225, 226, 243, 245
Adenyl kinase, 42, 43
ADP. *See* Adenosine diphosphate (ADP)
Aerobic metabolism, 39, 40
Aerobic phosphorylation, 39, 40
Aerobic respiration, 38, 39
Alaniz, Adam, 251
Alcoholic fermentation, 38, 46
Alway, Robert, *98,* 101–104, 115, 116
ALZA, 269, 270, 278, 279
Alzheimer's disease, 246
American Association for the Advancement of Science, 233
American Chemical Society, 38
American Medical Writers' Association, 241
American Philosophical Society, 125, 285
American Society of Biological Chemistry, 125
Ames, Bruce, 46
Amino acids, 44, 111, 122
Ammonium sulfate, 175
AMP. *See* Adenosine monophosphate (AMP)
Amyotropic lateral sclerosis, 7
Anaerobic fermentation, 38
Anaerobic metabolism, 39
Anaerobic respiration, 38, 39

Andreopoulos, Spyros, 162
Anheuser-Busch Inc., 63, 64
Animal cells
  pathway of glycogen breakdown in, 36, 37
  polyphosphates in, 226
  transphosphorylation reactions in, 43
Antibodies
  anti-idiotypic, 278
  catalytic, 142, 234
  genetically engineered, 272
  monoclonal, 274
Anti-Semitism, 8–11, 221
Aortic aneurysm, 7
Arai, Kenichi, 273, *275*, 275–77
Arai, Naoko, 273, 275–77, *276*
Arnold and Mabel Beckman Center, 99, 133, 148
Arnold and Mabel Beckman Foundation, 99
ARP. *See* Adenine ribose phosphate (ARP)
Arthritis, 267
Arthur (pet tiger), *82*
Arthur Kornberg Medical Research Building, 125
Asilomar, 150, 244
Aspartate transcarbamylase, 145
Association of American Medical Colleges, 113
Atlantic City, 62, 87
Atomic Bomb Casualty Commission, 204
ATP. *See* Adenosine triphosphate (ATP)
Australia, 167

Austro-Hungarian Army, 35, 36
Autonomous University of Barcelona, 30
Autonomous University of Madrid, 30
Ax, Emanuel, 181–183
Azobacter vinelandii, 81
Azotobacter vinelandii, 211

B
Bacillus subtilis, 159
Bacteria
  in DNA repair, 165, 166, 168
  in excision of pyrimidine dimers, 165, 166
  identifying by staining, 56
  organization of genetic material of, 20
  polyphosphate and, 226, 227
  RNA polymerase in, 107
  spores generated by, 194, 209
  swarming, 134, 135
  transforming factor, 88
  viruses, 159–61, 162, 197, 200, 212, 213, 226
Bacteriophages, 61, 86, 91, 111, 197, 200, 212
Baker, Tania, 138, 139, *238,* 239, 240, 248
Baldwin, William (Buzz), 108, 109, *109,* 131, 133, 135–137, 143, 152, 164
Bancroft Library, 9
Bangham, Alec, 209
Barker, Horace Albert (Nook), 71
Basel Institute of Immunology, 280
"Basic Research: the Lifeline of Medicine" (Kornberg), 263, 287

Bath Beach neighborhood in
  Brooklyn, 3
Bay Area, 79, 176, 179, 180, 182,
  183, 190, 239, 279
Beadle, George, 19, 20
Becker, Bernie, 10
Becking, Lourens Baas, 70
Beijerinck, Martinus, 70
Benjamini, Eliezer, 278
Benzer, Seymour, 111, 112, 151
Berg, Millie, 60
Berg, Paul, 4, 5, 46, 55, 59–61, 65,
  76, 102, 106, 108, 110, 111, 119,
  132, 133, *133*, 135, 139–41, 148,
  151, 154, 155, 175, 185, 218,
  220, 247, 248, 268, 272, 273,
  278, 280, 281, 285, 286–88
Berghof, The (Hitler's retreat home),
  61
Bernal, J. D., 209
Bessman, Maurice, 78, 86
Biochem. Biophys. Res. Comm.
  (journal), 185
Biochemica et Biophysica Acta
  (journal), 87
Biochemistry. *See also* Stanford
  Department of Biochemistry
    Cori laboratory and, 35–49, *37*
    Kornberg's curriculum in, 22, 58
    Kornberg's first publication in, 22
    Kornberg's introduction to, 7, 8,
      25, 26
    Ochoa laboratory and, 24–32, *30*
    Sylvy's Master's degree in, 23
Biotechnology Heritage Award, 271
*Biotech — The Countercultural
  Origins Of An Industry* (Vettel), 97

Biss, Paul, 181
Bloor, Walter R., 23
Board of Directors Meeting, Santa
  Monica, 232
Bodansky, Meyer, 8
Bohr, Aage N., 179n
Bohr, Niels, 179n
Bondy, Peter, 243, 244
Bonhoeffer, Friedrich, 170, 185
Book of Five Rings, A (Miamoto),
  241, 242
Borden Award, 113
Borden Foundation, 113
Boulder, Colorado, 108
Bower Award for Business
  Leadership, 270
Bragg, William Henry, 179n
Bragg, William Lawrence, 179n
Brain protein, 152, 153
Breathing, 213, 214
Brenner, Sydney, 169, 209
Bretscher, Marc, 92, 177
Bristol-Myers Company, 268
British Columbia, 76
British Columbia Research Council,
  112
Brooklyn, 3, 4, 8
Brooks, Mel, 215
Brutlag, Douglas (Doug), 149, 164,
  200, 201

C

CAD polypeptide, 145
Cairns, John, *167,* 167–70, 183–89
Cairns mutants, 166–70, 183–87
Cajal, Ramon, 24
Calcium, 147

California Medical Association, 124
California State Department of Education, 123, 124
"Call of Science The" (Cori, C.), 35
Caltech, 190, 278
Cambridge University, 152, 209
  Laboratory of Molecular Biology, 92, 169, 183, 271
Campbell, Peter, 243, 246
Cancer, 267
Candlestick Park, 76
Carbamyl phosphate synthetase, 145
Carbohydrates, 25, 38, 44, 150
Catalysis, 20, 21, 25, 27, 37, 38, 47, 93, 103, 121, 122, 166, 201, 210, 243, 288
Catalytic antibodies, 142, 234
Cathode-ray oscillograph, 63
Cell-free systems, 133, 206, 210, 213, 235
Cell membranes, 21, 226
"Centenary of the Birth of Modern Biochemistry" (Kornberg), 237
Center for the Advanced Study of Science and Technology, 11
Cetus Corporation, 268
Chain, Ernst, 263
Chance, Britton, 45
Chargaff, Edwin, 89n
Chase, Robert (Bob), 182
Chemical and Engineering News (magazine), 60
Chemical Basis of Heredity, The (Kornberg), 87
Chemistry
  organic, 28, 112, 175, 235
  physical, 28, 135, 150, 152

"Chemistry — the Lingua Franca of the Medical and Biological Sciences" (Kornberg), 236
Chester Beatty Research Institute, 152
Cholera, 17
Chromatin, 155, 177, 178
Chromosome replication, 161, 177, 199, 200, 206, 207, 213, 214, 217
Chromosomes, 161, 177, 199, 200, 206, 207, 213, 214, 217
"Chromosomes and Genes" (symposium presentation), 268
CIBA Lectures in Microbial Biochemistry (Kornberg), 237
Citrate synthase, 25
Citric acid (Krebs) cycle, 25, 27
City College of New York, 5, 6
Cleveland, Ohio, 60, 111, 144
Cleveland Clinic, 144, 146
Cleveland Orchestra, 180
Clinical Center, NIH, 53–55
Clinton, Bill, 265, 270
Cloning of genes, 92, 163n, 216, 217, 237, 274, 280
Clostridium birefringens, 193
Coenzymes, 20, 23–25, 40, 41, 43, 44, 64, 88, 242
Coffman, Bob, 278
Cohen, Stanley, 55, 152, 273
Cohn, Melvin, 59, 102, 106, 108, 112, 113, 132, 135, 142, 143, 149
Cold Spring Harbor Laboratory, 167
Colowick, Sidney, 42, 43
Columbia College, 181
Columbia University, 28, 170, 181, 183, 277, 278

Complementary DNA, 144, 145, 160, 161
Compton, Arthur, 54
Compton effect, 54
Conferences
  in Boulder, Colorado, 108
  Gordon Conference, 79
  in Italy, 235
Contagious diseases, 17
"Control of Gene Expression by Steroid Hormones" (symposium presentation), 268
Copenhagen, 60, 108
Cori, Carl, 4, 23, 25, 35–49, *37*, 53, 54, 56, 60, 63, 86, 102, 103, 105, 287
Cori, Gerty, 4, 25, 35–49, *37*, 54, 60, 63, 86, 287
Cornudella, L., 31
Corticodentatonigral disease, 246
Cosmos Club Award, 125
Cozzarelli, Nick, 160
Cranston, Alan, 11, 259
Creationism, 123, 124
"Creation of Life Rates Best of Science Stories in 1967" (newspaper article), 162
Creveling, James, 58
Crick, Francis, 69, 91, 92, 97, 119, 166, 169, 177, 187, 209, 287
Crucible of Science: The Story of the Cori Laboratory (Exton), 30
C-thymidine, 87, 212, 215
Curie, Pierre, 179n
"Cut in Basic Research Funds Called Tragic" (Kornberg), 261
Cyanide, 21

Cytidine triphosphate, 73
Cytochrome c, 20–22
"Cytochrome c-Cyanide Complex, The" (Kornberg), 22
Cytokines, 275, 280
Cytokinesis, 227
Cytosine, 71, 90

D

Dali, Salvador, 31, 32
Darwin, Charles, 124
Davis, Bernard, 13, 264
Davis, Ron, 149
DdPPK1 enzymes, 227
DdPPK2 enzymes, 227
DDT, 21
Dean's Selection Committee, 117
De Duve, Christian, 38, 55
Degenerative diseases, 234
Dehydrogenase
  isocitrate, 26, 31
  lactic, 39
  malate, 27
  oxidation of substance by action of, 59
  succinic, 20, 21, 31
De Kruif, Paul, 18
Delay, Dorothy, 181
DeLucia, Paula, 183
DeMars, Robert (Bob), 59, 62, 63, 106
De novo event, 201
Denver Post, The (newspaper), 260
Deoxyribonuclease (DNase), 85, 212, 215
Deoxyribonucleic acid (DNA). *See* DNA (deoxyribonucleic acid)

Department of Biochemistry at
  Kornberg as chair of, 5, 100–106
Deutscher, Murray, 164, 165
Dickson Prize in Medicine, 179
Dictyostelium disciodeum, 227
Digestion by pancreatic DNase, 85,
  212, 215
Dihydro-orotase, 145
Diphosphopyridine nucleotide
  pyrophosphatase, 41–43
Direct genetics, 214
Dirty enzymes, 196, 212
Diseases, 6, 7, 12, 17–19, 53, 56, 58,
  210, 228, 237, 255, 263, 265, 267
  degenerative, 234
  infectious, 17, 42, 280
  neurodegenerative, 246
  pellagra, 18, 19
  poverty and, 18
Dixon, Carolyn Frey. *See* Kornberg,
  Carolyn
Djerassi, Carl, 113, 175
DNA (deoxyribonucleic acid)
  complementary, 144, 145, 160,
    161
  containing pyrimidine dimers,
    165, 166
  electron microscopy of, 32, 141,
    149, 165, 177
  excision repair of, 165–67, 179
  repair, 165, 166, 168, 179, 237,
    273
  as template, 86, 87, 90–93, 122,
    160–62, 165, 178, 200, 206,
    207
  in vitro synthesis of, 87
DnaE, 186, 214

DNA-forming system, 88
DNA helicases, 207
DNA ligase, 133, 160, 207, 238
DNA polymerase
  discovery of, 31, 61, 64, 85–93,
    122
  E. coli, 159–61, 164–67, 170,
    186
DNA polymerase II, 170, 185, 186
DNA polymerase III holoenzyme,
  206, 207
DNA replication, 199–221
  replication fork, 200, 202, 203,
    207, 212, 213
  RNA priming during, 202, 203,
    205, 212
DNA Replication (Kornberg),
  238–240
DNA Replication, 2nd edition
  (Kornberg and Baker), 248
DNase. *See* Deoxyribonuclease
  (DNase)
DNA synthesis, 85–88, 141, 164,
  165, 170, 201, 207, 212, 214, 237
  in vitro, 85–87, 159, 170
DNA Synthesis (Kornberg), 237, 238
*DNAX Institute of Cellular and
  Molecular Biology, Inc. (DNAX),*
  249, 250, 272–81
DNAX–Schering–Plough
  partnership, 249, 250, 278–81
Doisy, Edward, 54
Dolly (cloned ewe), 163n
Dove, William (Bill), 201, 202
Drosophila melanogaster, 133
Drug delivery, 270
Dyer, Rolla, 14, 21

E

Eastern blotting, 145
Ecdysone, 133
E. coli
   DNA polymerase of, 90, 159–70, 186
   in enzymatic synthesis of RNA, 81, 82
Edsall, John, 89, 90
Electron microscopy, 32, 141, 149, 165, 177
Embden, Gustav, 25, 45
Embden–Meyerhof–Parnas pathway, 25, 45
Encyclopedia Britannica, 162
Engenics, 272
Enzymatic Synthesis of DNA (Kornberg), 89, 162, 237
Enzyme hunters, 236
Enzyme purification, 26–28, 38, 43, 85, 86, 89, 90, 164, 168, 196, 203, 205, 243
Enzymes
   catalysis, 20, 21, 25, 27, 37, 38, 47, 93, 103, 121, 122, 166, 201, 210, 243, 288
   dirty, 196, 212
   intracellular processes, 23, 24
   labile, 204
   "one gene, one enzyme" hypothesis, 20
   reversible synthesis, 45, 122
Enzyme Section at NIH, 41, 46
"Enzymic Synthesis of Deoxyribonucleic Acid," 87
Epidemics, 17
Erlanger, Joseph, 54, 63

Eukaryotes, 134, 155, 177, 179, 211, 214, 226
Excision repair of DNA, 165–67, 179
Exonuclease fragment, 169
Exonuclease function, 164–66, 169
Exton, John, 30

F

Falaschi, Arturo, 194, 195
Falwell, Jerry, 124
FASEB Journal, 252, 263
Fatty acids, 25
Federal funding for basic research, 233, 235, 255–65
Federation Meetings, 87
Federation of American Societies for Experimental Biology, 87, 263
Fire, Andrew, 119
Florey, Howard, 263
Flory, Paul, 113
Fluorescence spectroscopy, 147
Folic acid (vitamin B9), 18, 19
Förster resonance energy transfer (FRET), 147
For the Love of Enzymes: The Odyssey of a Biochemist (Kornberg), 3, 9, 28, 46, 163, 173, 193, 205, 219, 220, 233, 236, 239–43, 239–44, *241*, 287
   "Reflections on My Life in Science," 219, 220, 241, 287
Franklin Institute, 270
Free Academy of the City of New York, 5
FRET. *See* Förster resonance energy transfer (FRET)

Friedkin, Morris, 85
Fuller, Bob, 137, 206

G
Gairdner Foundation Award, 125
Galamian, Ivan, 181
Gasser, Herbert, 54, 63
Gefter, Malcolm, 183–89
Gellert, Marty, 160
Gene expression, 177, 268, 273
Gene hunters, 236
Genentech, 268
"General Model for Higher Organism Chromosomes, A" (Crick), 177
Genes, cloning of, 92, 163n, 216, 217, 237, 274, 280
Genetically engineered antibodies, 272
"Genetic Chemistry and the Future of Medicine" (Kornberg), 257
Genetic engineering, 163, 216, 217, 236, 241, 272, 287
Genetic recombination, 20, 110, 133, 237, 273
Genome, 159, 165, 177, 210, 211, 237
Genomics, 214
Germ Parade, The (Kornberg), 251, 252
Germ Stories (Kornberg), 251, 252
Gilbert, Augustin Nicolas, 12
Gilbert's Syndrome, 12, 14
Gladstone, Leonard, 107
Glaser, Robert (Bob) Joy, 115–18
Glucose, 25, 38, 39, 43, 45
Glycogen breakdown, 36–38

Glycolysis, 25, 45
GMP phosphodiesterase, 147
Goldberger, Joseph, 18
Goldberg–Hogness box (TATA box), 134
Golden Helix, The: Inside Biotech Ventures (Kornberg), 249–51, 269, 279
Goldstein, Avram, 99, 100, 103, 105, 106, 112, 114, 115
Gordon Conference, 79
Goulian, Mehran (Mickey), 160–64, *163*
Gray, Michael J., 230
Great Depression, 5
Green, David, 24
Gregory Pincus Award, 270
Griffith, Jack, *82,* 141
Grisiola, Santiago, 32
Grunberg-Manago, Marianne, 81, 121, 122
GTP-binding proteins, 275
Guanine, 73, 90
Guanosine 5′-diphosphate 3′-diphosphate (ppGpp), 226

H
"The hammer of enzyme purification" phrase, 89
Harris, Townsend, 5
Harvard College, 175, 176
Harvard School of Public Health, 167
Harvard University, 10, 89, 99, 101, 112, 143, 160, 217, 242, 250
    Cori appointed visiting professor of Biological Chemistry, 37

Department of Biological Chemistry, 175, 176
Medical School, 37, 116
School of Public Health, 167
Hass, Erwin, 20, 21
Hayaishi, Osamu, 58, 59, *70,* 70, 71
Healy, Bernadine, 263
Hemophilus, 159
Henry Kaplan and the Story of Hodgkin's Disease (Jacobs), 100, 114–18
Heppel, Leon, 20, 21, *22,* 41, 47, 54
Hershey, Al, 55
Hexokinase, 43
Higher organism, 154, 155, 211
Hilmoe, Russell, 21
Histones, 155, 177, 178
History of the World, Part 1 (movie), 215
Hoffman–LaRoche, Inc., 268
Hogness, David, 59, 60, *65,* 106, 108, 109, 123, 124, 133, 134, *134,* 135, 155
Hogness, Thorfin Rusten, 20
Holley, Robert, 112
Holmstrom, Kira, 229, 230
Honorary Fellow of the Weizmann Institute, 11
Hood, Leroy (Lee), 278
Horecker, Bernard (Bernie), 20–22, *22,* 24, 31, 41, 47, 54, 107
Hormones, 133, 134, 236
Hornig, Donald, 256
Horwitz, Rick, 168
Howard Hughes Investigatorship, 239

"How the 'Cancer War' is Wounding America" (Kornberg), 260, 261
Hughes, Sally Smith, 9, 10, 29, 46, 106, 110, 111, 124, 163, 186–89
  Berg interview, 280
  Kornberg interview, 9, 10, 29, 46, 106, 110, 111, 124, 163, 186–89, 217, 218, 225, 234, 258, 264, 267–69, 274
Humphrey, Hubert, 256
Hunkapiller, Michael, 278
Hurwitz, Jerard (Jerry), 59, *65,* 75, 106, 107, 122, 160, 187
Huxley, Thomas, 124
Hydroxymethyl group, 86
Hygienic Laboratory. *See* National Institutes of Health (NIH)
Hypertension, 267

I

Imperial Cancer Research Fund, 145–47, 167
Infectious diseases, 17, 42, 280
Inorganic polyphosphate, 214, 215, 225–227
Inorganic pyrophosphate (PPi), 40, 44
Institute of Microbiology, 237
Instituto Di Genetica, 189
International Symposium, 30, 31, 275
Interscience Publishers, 23
"Intestinal Menagerie" (Kornberg), 252
Introduction of Physiological Chemistry (Bodansky), 8
In vitro synthesis, 85, 86, 87, 159, 166, 169, 170, 201
Israel, 11, 81, 209, 210
Italy, 61, 189, 235, 238

J

Jackson Hole, 232
Jacob, Francois, 59, 87
Jacobs, Charlotte deCross, 100, 114
JAK-STAT pathway of signaling, 147
Japan, lecture in, 217
JBC. See Journal of Biological Chemistry (JBC)
JBC Online, 48
John Curtin School of Medical Research, 167
Johns Hopkins University, 10, 62, 87, 111
    Chemical Basis of Heredity symposium, 87, 88
    McCollum-Pratt Institute of, 87
Johnson, Lyndon, 71, 162, 256
Johnson, William, 113, 175
Joliot-Curie, Irene, 179n
Jones, Anne Fitz-Gerald. *See* Cori, Gerty
Josse, John, 91
Journal of Bacteriology, 210
Journal of Biological Chemistry (JBC), 22, 23, 30, 42, 45, 48, 89, 277
    JBC Online, 48
Journal of Clinical Investigation, 12
Julliard School, 170, 180, 181, 183, 186

K

Kafka, Franz, 36
Kaguni, Jon
Kaiser, Dale, 59, 60, 62, 63, *65,* 91, 106, 108, 109, 131, 133–35, *134,* 155, 245, 268
Kalckar, Herman, 23, 26, 39, 42, 45, 46, 60, 81
Kaplan, Henry, 99–104, 111–19, *115,* 151
Kaplan, Leah, 114, 118, 122
Karle, Jerome, 5
Katz, Lena. *See* Kornberg, Lena
Katzir, Ephraim, 11
Kaziro, Yoshito, 275–77
Keilin, David, 24
Kennedy, Edward (Ted), 262
Kennedy, Eugene, 242, 243
Kerr, Ian, 147
Khorana, Har Gobind, 76, 112
King's College, London, 87, 88
Kinyoun, Joseph J., 17
Klenow, Hans, 169
Klenow fragment, 169
Klug, Aaron, 176, 177
Knippers, Rolf, 170, 185
Korn, Edward, 46
Kornberg, Aaron Joseph, 1–3, *2,* 6
Kornberg, Arthur, *6, 13, 18, 65, 74, 78, 80, 123, 216, 286, 288*
    admission requirements for minority students, 219
    American Philosophical Society, elected to, 125
    Arthur Kornberg Medical Research Building dedicated to, 125
    authored books, 293
    bachelor degree from City College, 5, 6
    Baker's tribute to, 239
    Baldwin's tribute to, 131, 135
    Berg and Lehman's tribute to, 286, 287
    Berg's tribute to, 60, 61, 285

bilateral inguinal hernia repair, 45
Bretscher's tribute to, 92
British Columbia trip, 76
career highlights, 291, 292
as chair of Department
    Microbiology at Washington
    University, 53–60, 102, 175,
    193
as chair of Department of
    Biochemistry at Stanford
    University, 5, 100–106
as chief of *Enzyme Section* at
    NIH, 41, 46
as chief of Nutrition Laboratory
    at NIH, 18
childhood of, 1–6
co-founder, *DNAX,* 249
commercial ventures, 267–81
congressional testimonies on
    science funding, 257, 261
on creationism, 123, 124
death of, 60, 61, 77, 79, 92,
    131, 135, 137, 227, 231, 239,
    285–89, 286, 287
diaries of, 231, 232
discovers DNA polymerase, 31,
    61, 64, 85–93, 122
DNAX, 249, 250, 272–81
enters University of Rochester
    School of Medicine, 6–12
essay, editorials and other
    writings, 293–96
fall in bathroom, 285
Federation Meetings, 87
Foreign Fellow of Royal Society
    of London, elected as, 125
founding of *DNAX,* 249, 250,
    272–81

Gaspe Peninsula vacation, 43
grandparents, 1
"the hammer of enzyme
    purification" synonymous
    with, 89
honorary degrees, 32, 125,
    292
honorary degrees received, 292
as Honorary Fellow of the
    Weizmann Institute, 11
Hughes interview, 9, 10, 29, 46,
    106, 110, 111, 124, 163,
    186–89, 217, 218, 225, 234,
    258, 264, 267, 268, 269, 274
internship in Internal Medicine
    at Strong Memorial Hospital,
    7, 12, 21
intolerance of, 73, 77, 78, 113
Jackson Hole and Yellowstone
    National Park trip, 232
lecture in Jerusalem, 209, 210
Lederberg on Department of
    Biochemistry under Kornberg's
    leadership, 153
major awards, 292
memberships held by, 125
National Medal of Science, 71,
    125
as Nobel Laureate, 5, 32, 38,
    118–25, 179, 187, 244
notable honors, 125
at Ochoa's laboratory, 24–32
parents, 1–7, *2, 6, 18,* 193
Paul Lewis Award, 45, 46, 60
postdoctoral training with Carl
    and Gerty Cori, 35–49
postdoctoral training with
    Ochoa, 24–32

as president of American Society of Biological Chemistry, 125
religious prejudice encountered as Jew, 8, 9
relinquishes chairmanship of Stanford Department of Biochemistry, 217
Schekman's obituary of, 77
scholarship from University of Rochester Medical Center patient, 8
70th birthday celebration, 77, 78, 287
Soviet Union tour, 118
Stanford Medical School's tribute to, 79
Strong Memorial Hospital internship, 7, 12, 21
United States Navy and, 12–14
Villa Serboloni trip, 238
viral DNA, first replication of, 159, 162
vitamins and, early interest in, 23
White House invitation to meet President Johnson, 256
wife Carolyn, 248, 249, 285
wife Charlene, 248, 285
wife Sylvy, 7, 23, 26, 28–30, 43, 46, *47,* 48, 57, 60, *74,* 86, 101, 113, 121, 122, 173, 176, 179, 221, 225, 238, 239, 240, 243–48, *246,* 287
Kornberg, Bella, 1
Kornberg, Carolyn, 248, 249, 285
Kornberg, David Lieb, 1
Kornberg, Ella, 3
Kornberg, Jody, 189

Kornberg, Ken, 23, *74,* 78, *140,* 173, 176, 179, 186, *189,* 189, 190, 244, 245, 247, 248, 274, 289
Kornberg, Lena, 2, *2, 6, 7,* 193
Kornberg, Martin, 3, 4
Kornberg, Roger, 23, 54, *74,* 75–80, 114, 115, 118, 119, 120, 139, 140, *140,* 143, 149, 153–54, 155, 173–79, *178,* 183, 186–88, 209, 229, 231, 232, 233, 244, 245, 247, 248, 278, 285, 287–89
Kornberg, Ross, 189
Kornberg, Sylvy Ruth, 7, 23, 26, 28–30, 43, 46, *47,* 48, 57, 60, *74,* 86, 101, 113, 121, 122, 173, 176, 179, 221, 225, 238, 239, 240, 243–48, *246,* 287
Kornberg, Tom, 41, *74,* 76, 78, *140,* 170, 173, 179–89, *180,* 229, 243, 244, 245, 247, 278, 286
Kornberg, Veronica, 190
Kornberg, Zac, 189
Kornberg Associates, 190
Kornberg Journal Club, 232
Koshland, Daniel (Dan), 135, 288, 289
Krebs, Edwin G., 38, 54
Krebs citric acid cycle, 25, 27
Krell, Bella. *See* Kornberg, Bella
Kunitz, Moses, 85

L

Laboratory of Molecular Biology (LMB), 92, 169, 183, 271
Lactic acids, 39
Lardy, Henry, 45

"Latent Liver Disease in Persons Recovering From Catarrhal Jaundice and in Otherwise Normal Medical Students, as Revealed by the Bilirubin Excretion Test" (Kornberg), 12
Lateral diffusion, 176
Lectures/seminars
　affiliate, 268
　on isolation of enzymes from cellular juices of microbes, 57
　Italy, 1988, 235
　Japan, "DNA Replication From Start to Finish," 217
　Jerusalem, 2000, 209, 210
　Kornberg as visiting seminar speaker at NIH, 120
　Kornberg's critique of students presenting, 140, 141, 144
　Kornberg's preparation for, 231
　at NIH presented by Tatum, 19, 20, 99
　"Regulation of Cellular Calcium by Yeast Vacuoles," 232
　Tabor lunch seminars, 48
　"Two Cultures, The: Chemistry and Biology" at American Association for the Advancement of Science, 1987, 233
　University of Rochester Medical School, 1997, 9
　at University of Wisconsin, 225
Lederberg, Joshua, 10, 20, 109–11, *111,* 153, 196, 242
Lee, Frank, 278
Lee, Jung, 217

Lehman, Robert (Bob), 59–62, *62, 65,* 77, 86, 106, 133, 135, 147, 149, 160, 164, 186, 245–47, 286
Lehninger, Albert, 10, 45
Leloir, Luis F., 38, 55
Lerner Research Institute of the Cleveland Clinic, 144
Levering, Charlene, 237, 240, 241, 248, 285
Levi-Montalcini, Rita, 55
Levy, Ronald, 278
Levy, Sylvy Ruth. *See* Kornberg, Sylvy Ruth
Libby, Willard F., 112n
Lieberman, Irving, 58, 62, 71–73
Lindberg, Olov, 40, 42
Lipids, 7, 23, 31, 44, 64, 145, 150
Lipmann, Fritz, 23, 39, 139
Liposomes, 209
Littauer, Uri, 81
LMB. *See* Laboratory of Molecular Biology (LMB); Medical Research Council Laboratory of Molecular Biology (LMB)
Lobban, Peter, 131
Loewi, Otto, 36
Lohmann, Karl, 39
London Times (newspaper), 163
Lorch, Yahli, 178
Loring, Hubert, 97, 99, 104
Los Angeles Times (newspaper), 261
Lowry, Oliver, 42
Luciano, Bob, 250, 279
Luck, Murray, 97, 99, 104
Luria, Robert (Bob), 76
Luria, Salvador, 59

Lwoff, Andre, 62, 63
Lymphokines, 280
Lymphoma, 7
Lynen, Feodor, 80, 139

M
Ma, Yo-Yo, 181
Macarthur, Donald M., 256
Magnetic resonance imaging (MRI), 265
Magnetic resonance studies, Roger Kornberg's, 209
Mahler, Alan, 27
Malate dehydrogenase, 27
Marine Biological Station, 35
Marine Hospital Service (MHS), 17
Markham, Roy, 21
Marshall Plan, 11
Massachusetts Institute of Technology, The, 138
Matthaei, J. Heinrich, 122
Max Planck Institute, 170
McArdle Laboratory, 201
McCann, William S., 8
McCloskey, Paul N., 261
McCollum-Pratt Institute of Johns Hopkins University, 87
McConnell, Harden, 113, 176, 209, 278
Medical College of Georgia, 174, 175
Medical Corps of the US Army and Navy, 12, 13
Medical Research Council Laboratory of Molecular Biology (LMB), 176, 177
Medical Service Plan, 116
Mendel, Gregor, 259

Merck & Co, 281
Messenger RNA (mRNA), 122
Metabolism, 7, 20, 23–25, 31, 37–40, 44, 45, 235, 237, 238
Meyerhof, Otto, 23, 25, 45, 287
MHS. *See* Marine Hospital Service (MHS)
Miamoto, Musashi, 241, 242
Michaels, Alan, 272
Microbe hunters, 236
Microbe Hunters (de Kruif), 18
Mill Hill Laboratory, 167
Mitchell, Peter, 209
Mitochondria, 39, 40, 47, 238
Model DU spectrophotometer, 21, 22
Model organisms, 154, 155
Molecular biology, 24, 148, 154, 173, 183, 193, 195, 196, 215, 233–36, 240, 241, 261, 271–274, 287
Molecular Cell (journal), 230
Molecular fossil, 225, 226, 228
Molteno Institute, 21
Monoclonal antibodies, 274
Monod, Jacques, 59
Moore, Kevin, 278
Mossman, Tim, 278
MRI. *See* Magnetic resonance imaging (MRI)
MRNA. *See* Messenger RNA (mRNA)
M13 bacteriophage, 200–202, 208, 209, 212, 213
"Multifarious Molecules" (*Nature* editorial), 169
Mutagenesis, 164

Mutants
Mutants, Cairns, 166–70, 183–87
"My Faith in Science" (Kornberg), 4
Myokinase, 42, 43
Myxococcus xanthus, 134, 135

N
NAD. *See* Nicotinamide adenine dinucleotide (NAD)
NADH. *See* Nicotinamide adenine dinucleotide, reduced form (NADH)
NADP. *See* Nicotinamide adenine dinucleotide phosphate (NADP)
Nathans, Daniel, 10, 55
National Academy of Sciences, 125
National Cancer Institute, 7, 23
National Health Research Fellowship and Traineeship Act of 1973, 259, 260
National Historic Chemical Landmark, 38
National Institutes of Health (NIH), 13, 14, 17–32, 41–49, 53–56, 58–60, 62, 64, 70–73, 99, 120, 160, 173, 189, 240, 242, 257, 258–60, 262, 263
    budget for basic science, 257, 262
    Clinical Center, 53–55
    Enzyme Section, 41, 46
    Nutrition Laboratory, 18
National Medal of Science, 71, 125
National Medal of Technology, 270
National Science Foundation (NSF), 257, 259, 262
Nature (journal), 75, 168–70, 183–85, 243, 264

Nature Communications (journal), 230
Nature New Biology (journal), 170, 187, 212
Nearest-neighbor analyses, 91, 92, 160
Negrin, Juan, 24
NER. *See* Nucleotide excision repair (NER)
Nerve growth factor, 152
Neurodegenerative disease, 246
Neurofibrillary tangles, 246
New England Journal of Medicine, 250, 251
New York, Kornberg and
    childhood, 1–6, *6*
    medical school, 5–9
    social life with Sylvy, 29, 30
New York Philharmonic Orchestra, 180
New York University School of Medicine, 107
Nick translation, 165, 166
Nicotinamide adenine dinucleotide (NAD), 20, 21, 40, 41, 43–45, 242, 243
Nicotinamide adenine dinucleotide, reduced form (NADH), 25
Nicotinamide adenine dinucleotide phosphate (NADP), 20, 21, 43, 45
Nicotinamide-ribose phosphate (NRP), 41, 43, 44, 73, 243
NIH. *See* National Institutes of Health (NIH)
Nirenberg, Marshall, 112, 122
Nixon, Richard, 259–61
Nobel Laureates

from Abraham Lincoln High School, 5
Beadle, George, 20
Berg, Paul, 5, 119
Cajal, Ramon, 24
Compton, Arthur, 54
Cori, Carl and Gerty, 37, 38
de Duve, Christian, 38
Doisy, Edward, 54
Erlanger, Joseph, 54, 63
father and daughter duo, 179n
father/son teams, 179n
Fire, Andrew, 119
Gasser, Herbert, 54, 63
Karle, Jerome, 5
Klug, Aaron, 177
Kornberg, Arthur, 5, 32, 38, 118–25, 179, 187, 244
Kornberg, Roger, 119, 177, 179
Krebs, Edwin G., 38
Leloir, Luis F., 38
Loewi, Otto, 36
Meyerhof, Otto, 25, 45
Ochoa, Severo, 32, 38, 118, 119, 121, 122
Rothman, James, 149
Sakharov, Andrei, 123
Schekman, Randy, 77, 136
from Stanford Medical School, 119
Sudhof, Tom, 119
Sutherland, Earl W., 38, 55
Tatum, Edward, 20
from Washington University, St. Louis, 38, 54, 55
Whipple, George, 9

Nobel Prize
Goldberger nominated for, 18
Kornberg on being a Nobel Laureate, 122, 123
Kornberg on nomination process, 119, 120
Kornberg's acceptance speech, 121, *123*
Kornberg's request to permit Sakharov to receive, 123
Nobel Prize Foundation, 118, 121. *See also* Nobel Laureates; Nobel Prize
N-phosphonacetyl-L-aspartate (PALA), 145
NRP. *See* Nicotinamide-ribose phosphate (NRP)
NSF. *See* National Science Foundation (NSF)
Nuclear resonance techniques, 176
Nucleic acid enzymology, 69–82
Nucleosides, 44, 81, 85, 194, 243
Nucleosomes, 177, 178
Nucleotide excision repair (NER), 166, 178, 179
Nucleotides, 30, 41–45, 64, 69, 70, 73, 88, 93, 111, 122, 145, 160, 161, 164, 243

O

Ochoa, Carmen, 32
Ochoa, Severo, 24–32, *30,* 38, 39, 40, 55, 73, 80–82, 119, 121, 122, 196, 197, 211, 275
Officers Service Club, 26
Okazaki, Reiji, 203–5
Okazaki courage, 204

Okazaki fragments, 203, 207
Okazaki maneuver, 203
Olins, Don and Ada, 177, 178
"One gene, one enzyme" hypothesis, 20
Organic chemistry, 28, 112, 175, 235
OriC, 205, 206, 213, 214
Oro, J., 31
"Osamu Hayaishi: Pioneer of the Oxygenases, the Molecular Basis of Sleep and Throughout a Great Statesman of Science" (Kornberg), 59
Overgaard-Hansen, Jørgen, 169
Oxygenases, 59

P
Pacific Grove, 57
Pacific Grove, California, 57, 244
PALA. *See* N-phosphonacetyl L-aspartate (PALA)
Palo Alto, 97–99, 102, 116, 248, 269, 270
Palo Alto City Council, 98
Palo Alto-Stanford Medical Center, 99
Pancreatic DNase, 85, 215
Pantothenic acid (vitamin B5), 18
Paramagnetic resonance techniques, 176
Parnas, Jakob Karol, 25, 45
Pasteur Institute, 59, 62
"Pathways of Enzymatic Synthesis of Nucleotides and Polynucleotides" (Kornberg), 3, 9, 28, 46, 88, 163, 173, 193, 205, 219, 233, 236

Paton, Noel, 24
Paul Lewis Award, 45, 46, 60
Pellagra, 18, 19
Penicillin, 263
Pentose phosphate pathway, 21
Perutz, Max, 163, 271
Pfizer Award. *See* Paul Lewis Award
Pharmaceutical companies, 247, 267, 268, 270, 279
Phenylalanine, 122
ΦX174 DNA, 159–61, 162, 212, 213
Phosphate, 7, 39, 40, 41, 43, 44, 69, 73, 215, 226
Phosphoanhydride bonds, 225
Phosphodiesterase, 21, 147
Phospholipid flip-flop, 176
Phospholipids, 7, 44, 69
Photosynthesis, 25, 57n
PHS. *See* US Public Health Service (PHS)
Physical chemistry, 28, 135, 150, 152
Plasmids, 152, 206, 211, 238
PNAS, 89, 185, 186, 200, 201
Pneumococcus, 159, 215
Podgorny, Nikolai V., 123
Poliovirus, 252
Polyamines, 47
Polynucleotide phosphorylase, 81, 82, 121, 211
PolyP. *See* Polyphosphate (PolyP)
Polyphosphate (PolyP), 225–30
"Polyphosphate is a Primordail Chaparone" (Gray), 230
Polysaccharides, 43
"Possible Role for RNA Polymerase in the Initiation of M13 DNA

Synthesis, A" (Kornberg, Brutlag, and Schekman), 201
Potato nucleotide pyrophosphatase, 43
Potter, Van, 45
Pound, Ezra, 251
PPi. *See* Inorganic pyrophosphate (PPi)
PPK1 enzymes, 227
PPK2 enzymes, 227
President's Daily Diary Worksheet, 256
Pricer, Bill, 47
Primosome, 206–8, 212
Prize in Medicine or Physiology, 119
Proceedings of the American Philosophical Society, 285
Protein, 27, 39, 150, 169, 177, 185, 189, 197, 199, 211, 213, 214, 227, 234, 236, 274, 280
   amino acids and, 44, 111
   brain, 152, 153
   concentration of, Lowry's assay for determining, 42
   for DNA replication, 64, 205–7, 209, 214, 216, 217
   for priming, 205
   in regulation of transcription in living cells, 178
   synthesis, 31, 44, 47, 64, 133, 205, 272, 275
   tau, 246
   transfer, 145
   transverse diffusion, 176n
   turnover, 194
Proteolipid hypotheses, 209
Public Health Service Act, 259
PubMed, 209
Purine, 70, 73, 194
Pyrimidine dimers, 165, 166
Pyruvate, 45

R

Racker, Efraim, 212
Radiolabeled thymidine, 85, 88
Reagents, 8, 41, 86, 132, 214, 215, 234, 273
Receptor, 236, 280
Recombinant DNA technology, 133, 163, 215, 236, 240, 241, 261, 268, 272, 273, 287, 288
Reflections on Biochemistry — In Honour of Severo Ochoa (Kornberg, Horecker, Cornudella, and Oro), 31
"Regulation of Cellular Calcium by Yeast Vacuoles" (seminar), 232
Reichard, Peter, 119
Renick, Donna, 278
Replication, 199–221
Replication fork, 200, 202, 203, 207, 212, 213
Replisome, 206–8, 240
Research funding, 233, 235, 255–65
Reverse genetics, 214
Riboflavin (vitamin B2), 18
Ribonucleic acid (RNA). *See* RNA (ribonucleic acid)
Richardson, Charles, 160, 164, 175, 186, 187
Rifampicin, 201, 208
RNA (ribonucleic acid)
   messenger (mRNA), 122

priming during DNA
 replication, 202, 203, 205, 212
 synthesis, 85, 201, 202
 transfer (tRNA), 112, 132
Robertson, Channing, 272
Roche Institute of Molecular
 Biology, 30
Rockefeller Foundation, 238
Rose, Leonard, 180, 181, 186
Roslin Institute, 163n
Roswell Park Cancer Institute, 36
Rothman, James, 148, 149
Rowen, Lee, 137, 138
Royal Society of London, 125

S

Saccharides, 31
St. Louis Cardinals, 76
St. Louis Walk of Fame, 37
Sakharov, Andrei, 123
Salk Institute, 143
San Francisco 49ers, 276
San Francisco Giants, 76
San Francisco Symphony Orchestra, 180, 182
Schekman, Randy, 77, 136, 200–202
Schering–Plough Corporation, 249, 250, 278–81
School of Microbiology at the John Curtin School of Medical Research, 167
SCIENCE (journal), 255, 262, 263
Scopes, John, 124
Scripps Institution of Oceanography, 189
Sebrell, William Henry, 18, 19, 23, 41

Sekimizo, Kazuhisa, 217
Seminars. See Lectures/seminars
Shannon, James, 257, 263
Shannon Awards for Investigator-Initiated Research, 263
Shooter, Eric, 152, 153, 155, 246
Siegbahn, Kai M., 179n
Siegbahn, Manne, 179n
Signaling, JAK-STAT pathway of, 147
"Signaling Properties of Inorganic Polyphosphate in the Mammalian Brain" (Holmstrom), 229, 230
Simms, Ernie, 73, 75, 86, 105, 175
Sinsheimer, Robert (Bob), 159, 160, 162
Smith, Hamilton, 55
Smith, John D., 21
Smith Kline & French Laboratories, 268
Smithsonian Institute, 162
Snell, George, 55
Southern, Edward (Ed), 145
Southern blotting, 144, 145
Spectrophotometry, 21, 22, 26
Spudich, James (Jim), 194–96, *195*, 245
Stanford Department of Biochemistry, 131–55, 194
 Arai awarded consulting professorship, 277
 four-hour "teach-in" in memory of Kornberg, 285
 Kornberg as chair of, 5, 100–106
 Lederberg on Kornberg's leadership of, 153
Stanford Industrial Park, 274

Stanford University
   Architectural Design Program, 190n
   Black Student Union, 219
   Department of Architecture, 190
   Department of Biochemistry (*See* Stanford Department of Biochemistry)
   Department of Biology, 99, 111, 278
   Department of Chemical Engineering, 272
   Department of Chemistry, 97, 104, 112, 113, 154, 175, 176, 268
   Department of Electrical Engineering, 218
   Department of Genetics, 110, 151–53, 196, 242
   Department of Neurobiology, 153
   Department of Pathology, 278
   Department of Pharmacology, 99
   Department of Radiology, 99, 114
   Department of Structural Biology, 178
   Medical Affairs at, 117
   Medical School Faculty Senate, 219, 221
   Oceanographic Institute, 57
   Palo Alto-Stanford Medical Center, 99
   racial tension, 218–21
   School of Medicine, 79, 97, 104, 110, 119, 154, 218
   Stem Cell Center, 278
   University Board of Trustees, 97
Stark, George, 143–46, *146,* 149
State Institute for the Study of Malignant Disease, 36
Stein, Jay, 9
Sterling, Wallace, 97, 98, 102, 104, 116, 117
Stone, Edward Durrell, 99, 101
Storm Over Biology: Essays on Science, Sentiment, and Public Policy (Davis), 264
Stratling, Wolf, 170
Strong Memorial Hospital, 7, 12, 21
Stryer, Lubert, 143, 144, *146,* 147–49
Succinoxidase, 21
Sucrose, synthesis of, 71
Sudhof, Tom, 119
Sugars, 38, 145
Sutherland, Earl W., 38, 55
Swarming, 134, 135
Symposium presentations
   Chemical Basis of Heredity, The, 87
   "Chemistry — the Lingua Franca of the Medical and Biological Sciences," 236
   "Chromosomes and Genes," 268
   "Control of Gene Expression by Steroid Hormones," 268
   in Houston, Texas, 175
   for industrial affiliates, 268, 269
   International Symposium in Madrid, 30, 31, 275
   at McCollum-Pratt Institute, 87, 88

"Pathways of Enzymatic Synthesis of Nucleotides and Polynucleotides," 3, 9, 28, 46, 88, 163, 173, 193, 205, 219, 233, 236
  at Rockefeller University, 236
"Targeting of Proteins Into Eukaryotic Cells," 268
Today's Opportunities, Tomorrow's Health: The Future of Biomedical Research in America, 263
Syntex Corporation, 270
Syphilis, 58

T
Tabor, Herbert (Herb), 47–49, *49*
"Targeting of Proteins Into Eukaryotic Cells" (symposium presentation), 268
TATA box (Goldberg–Hogness box), 134
Tatum, Edward (Ed), 19, 20, 99
Tau, 246
Taube, Henry, 113
Taylor, John, 39
T cells, 280
Temple-Black, Shirley, 98
"Ten Commandments: Lessons From the Enzymology of DNA Replication" (Kornberg), 210–15
Terman, Frederick E., 97, 100–102, 104, 105
TFIIH. *See* Transcription factor IIH (TFIIH)
Thiamin (vitamin B1), 18
3rd Infantry Division, 61

Thomson, George Paget, 179n
Thomson, J. J., 179n
Thymidine
  C-thymidine, 87, 212, 215
  radiolabeled, 85, 88
Thymine, 71, 90
Thymine-thymine dimers, 166
TIBS. *See Trends in Biochemical Sciences* (TIBS)
Today's Opportunities, Tomorrow's Health: The Future of Biomedical Research in America (Kornberg), 263
Tomkins, Gordon, 46
Transcription, 134, 155, 178, 179, 201, 202, 213, 214, 237
Transcription factor IIH (TFIIH), 178, 179
Transfer RNA (tRNA), 112, 132
Transverse diffusion, 176n
Trends in Biochemical Sciences (TIBS), 215
Tribute to Arthur Kornberg, MD (Stanford Medical School), 79
TRNA. *See* Transfer RNA (tRNA)
T2 bacteriophage, 86, 91
Tumor-virus genes, 272
"Two Cultures, The: Chemistry and Biology" (Kornberg), 233

U
Ultraviolet (UV) radiation, 165, 166
US Army and Navy, Medical Corps of, 12–14
US Congress, 17, 257, 258, 262
US Marine Hospital, 45
US Public Health Service (PHS), 17, 18

US Senate Subcommittee on Government Research, 257
University of British Columbia, 76
University of California at Berkeley, 11, 71, 100, 101, 105, 278, 288
University of California at San Francisco, 9, 173, 190
University of Colorado School of Medicine, 116
University of Graz, 36
University of Houston, 30
University of Oregon, Department of Neurobiology at, 153
University of Pavia, 189
University of Rochester Medical Center, 7, 8, 9, 11, 12, 125
University of Texas Southwestern Medical Center, 119
University of Wisconsin, 19, 108, 109, 201, 225, 239
Upjohn Company, 268
Uracil, 71–73
UV radiation. *See* Ultraviolet (UV) radiation

V

Valenti, Jack, 256
Van Niel, Cornelius Bernardus (C.B.), 57, 71
Van Tamelen, Eugene, 113
Vettel, Eric, 97, 98
Vietnam War, 218, 256
Villa Serboloni, 238
Viral DNA, 159, 162, 291
Viral infection, 196, 197, 208
Viruses, 32, 47, 159–62, 197, 200, 210, 212, 213, 238, 252, 272

Virus of Anti-Semitism, The (Kornberg), 287
Virus Research Institute, 167
Vitamin B, 25
Vitamin B1 (thiamin), 18
Vitamin B2 (riboflavin), 18
Vitamin B5 (pantothenic acid), 18
Vitamin B9 (folic acid), 18, 19
Vitamin B12, 71
Vitamin hunters, 18, 19, 236
Vitamin K, 54
Vitamins, 18–20, 23, 25, 54, 71, 251, 252
Von Euler, Ulf, 179n
Von Euler-Chelpin, Hans, 179n

W

Waitz, Allan, 279–81
Walsh, Bill, 276
Walter and Eliza Hall Institute of Medical Research, 167
Warburg, Otto, 20, 21, 23, 41, 287
War on Cancer (Nixon's declaration of), 259–61
Washington University, St. Louis, 107, 108, 110, 112, 113, 131, 136, 139, 140, 142, 143, 151, 154, 193, 217, 225, 244, 245, 286
  Department of Microbiology, 53–60, 102, 175, 193
  Department of Pharmacology, 35, 36, 39, 42, 85
  Executive Committee, 55, 101, 102, 115–17

School of Medicine, 36, 38, 42, 53, 58, 103
Watson, James, 69, 91, 97
Watson–Crick paper, 69, 91, 97
Webster, Horace, 5
Weinberger, Casper, 260, 261
Weismann, Irving, 278
Weiss, Samuel, 107
Weizmann Institute, 11, 81
Western Reserve University, Department of Biochemistry at, 60
W. H. Freeman & Co, 237
Whipple, George, 9, 11, 12
Whipple Auditorium, 9
White, Celia, 47
White House invitation, 256
Wickner, Sue, 106
Wickner, William (Bill), 77, 106, 141
Wilkins, Maurice, 87, 88
Wolfe, Sophie, 4
Wood, Barry, 10
Wood, Harland, 60
Worcester Foundation, 270

World War I, 1, 35
World War II, 12, 36, 61, 204, 257, 262

X
XOMA, 232
X ray diffraction, 88

Y
Yanofsky, Charles, 111, 175, 272, 273, *273*, 278
Yasuda, Seichi, 213
Yeast cells, 38, 71
Yellow fever, 17
Yellowstone National Park, 232

Z
Zaffaroni, Alejandro (Alex), 247–50, 269–71, *271*, 272, 273, 274, 278, 279, 281
Zimmerman, Steven, 86
Zurawski, Gerard, 278
Zymobacterium kornbergii, 71
Zymobacterium oroticum, 71

www.ingramcontent.com/pod-product-compliance
Lightning Source LLC
Chambersburg PA
CBHW052045220426
43663CB00012B/2444